▶   **Russia and Latin America**

DOI: 10.1057/9781137308139

Other Palgrave Pivot titles

Ramin Jahanbegloo: **Democracy in Iran**

Mark Chou: **Theorising Democide: Why and How Democracies Fail**

David Levine: **Pathology of the Capitalist Spirit: An Essay on Greed, Hope, and Loss**

G. Douglas Atkins: **Alexander Pope's Catholic Vision: "Slave to No Sect"**

Frank Furedi: **Moral Crusades in an Age of Mistrust: The Jimmy Savile Scandal**

Edward J. Carvalho: **Puerto Rico Is in the Heart: Emigration, Labor, and Politics in the Life and Work of Frank Espada**

Peter Taylor-Gooby: **The Double Crisis of the Welfare State and What We Can Do About It**

Clayton D. Drinko: **Theatrical Improvisation, Consciousness, and Cognition**

Robert T. Tally Jr.: **Utopia in the Age of Globalization: Space, Representation, and the World System**

Benno Torgler and Marco Piatti: **A Century of** American Economic Review: **Insights on Critical Factors in Journal Publishing**

Asha Sen: **Postcolonial Yearning: Reshaping Spiritual and Secular Discourses in Contemporary Literature**

Maria-Ionela Neagu: **Decoding Political Discourse: Conceptual Metaphors and Argumentation**

Ralf Emmers: **Resource Management and Contested Territories in East Asia**

Peter Conn: **Adoption: A Brief Social and Cultural History**

Niranjan Ramakrishnan: **Reading Gandhi in the Twenty-First Century**

Joel Gwynne: **Erotic Memoirs and Postfeminism: The Politics of Pleasure**

Ira Nadel: **Modernism's Second Act: A Cultural Narrative**

Andy Sumner and Richard Mallett: **The Future of Foreign Aid: Development Cooperation and the New Geography of Global Poverty**

Tariq Mukhimer: **Hamas Rule in Gaza: Human Rights under Constraint**

Khen Lampert: **Meritocratic Education and Social Worthlessness**

G. Douglas Atkins: **Swift's Satires on Modernism: Battlegrounds of Reading and Writing**

David Schultz: **American Politics in the Age of Ignorance: Why Lawmakers Choose Belief over Research**

G. Douglas Atkins: **T.S. Eliot Materialized: Literal Meaning and Embodied Truth**

Martin Barker: **Live To Your Local Cinema: The Remarkable Rise of Livecasting**

Michael Bennett: **Narrating the Past through Theatre: Four Crucial Texts**

Arthur Asa Berger: **Media, Myth, and Society**

Hamid Dabashi: **Being a Muslim in the World**

David Elliott: **Fukushima: Impacts and Implications**

Milton J. Esman: **The Emerging American Garrison State**

Kelly Forrest: **Moments, Attachment and Formations of Selfhood: Dancing with Now**

Steve Fuller: **Preparing for Life in Humanity 2.0**

Ioannis N. Grigoriadis: **Instilling Religion in Greek and Turkish Nationalism: A "Sacred Synthesis"**

DOI: 10.1057/9781137308139

palgrave▶pivot

Russia and Latin America: From Nation-State to Society of States

Marvin L. Astrada and
Félix E. Martín

DOI: 10.1057/9781137308139

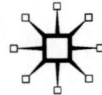

First published in 2013 by
PALGRAVE MACMILLAN®
in the United States—a division of St. Martin's Press LLC,
175 Fifth Avenue, New York, NY 10010.

Where this book is distributed in the UK, Europe and the rest of the
world, this is by Palgrave Macmillan, a division of Macmillan Publishers
Limited, registered in England, company number 785998, of Houndmills,
Basingstoke, Hampshire RG21 6XS.

Palgrave Macmillan is the global academic imprint of the above
companies and has companies and representatives throughout the world.

Palgrave® and Macmillan® are registered trademarks in the United States,
the United Kingdom, Europe and other countries.

ISBN: 978-1-137-30814-6 EPUB
ISBN: 978-1-137-30813-9 PDF
ISBN: 978-1-137-30812-2 Hardback

Library of Congress Cataloging-in-Publication Data is available from the
Library of Congress.

A catalogue record of the book is available from the British Library.

First edition: 2013

www.palgrave.com/pivot

DOI: 10.1057/9781137308139

Marvin L. Astrada

I dedicate this book to Alexandra Astrada and Jessica C. Pino; the two women in my life who I love dearly, and who have continued to provide unwavering support for my endeavors—professional and otherwise. First, I dedicate this book to my mother, Alexandra Astrada, who remains a constant and powerful source of inspiration and joy in my life. Her devotion and passion for exploring and engaging life—for questioning, explaining, and understanding the deep complexity that underlies human affairs—remains the standard I follow in all aspects of my life. Second, I dedicate this book to Jessica C. Pino, who has proven to be a wonderful friend, companion, and colleague. I thank her for the countless, enjoyable, and ongoing discussions, debates, queries, analyses, and critiques we have shared that span the gamut of issues to be found in politics and international relations—all of which have found expression in this book.

Félix E. Martín

I dedicate my work in this book to present and former graduate students, like Marvin Astrada, my coauthor and dear friend, who continue to impress me with their laborious dedication, inquisitive zeal, and noble efforts to further our understanding of international politics in an ever more increasing complex world. To them all I extend my most sincere and heartfelt dedication.

DOI: 10.1057/9781137308139

Contents

Preface   viii

Introduction   1

1   Mapping Complex Cooperative Networks:
    Society in International Organization   10
        Society, complexity, networks and
            Russia–Latin America relations   12
        A systems approach to international
            society and organization   14
        Agents, CCNs, and international
            organizations   18
        Conclusion   21

2   Globalization, International Organization
    and the Rise of a Society of States   26
        CCNs: adaptive learning and
            international order   36
        Components of CCNs: dynamics
            and implications   40
        Challenging the system of
            states paradigm   42
        System and society:
            utterly antithetical paradigms?   45

3   Exploring the Emergent State-Society
    Synthesis: Russia–Latin America Relations   57
        The Cold War, the USSR, and Russia   62

DOI: 10.1057/9781137308139

Soviet political ideology and foreign policy objectives   63

Russian foreign policy and Latin America   67

4   Building Complexity: Select Case Studies of
CCNs—Russia and Latin America   76

    Mexico   78

    Colombia   81

    Venezuela   84

    Cuba   85

    Nicaragua   87

    Bolivia   90

    Brazil   92

    Ecuador   93

5   Concluding Thoughts on CCNs in
Russia–Latin America Relations   101

Bibliography   110

Index   128

DOI: 10.1057/9781137308139

Preface

This work explores alternative international mechanisms and principles of order and organization; i.e., it explores and analyzes the emergence of complex cooperative networks in interstate relations. Networks present challenges and opportunities for states residing in an international system of states permeated by *realpolitik*. Networks provide alternative bases for interstate engagement based on cooperation. This is the case because networks, as a tool of statecraft, have substantively impacted the dynamics of international order and organization by integrating states to an unprecedented degree. The networking of states' interests, purpose, and wellbeing, e.g., dissemination and acceptance of the notion of there being universal and prosecutable "crimes against humanity," is characterized by the fusion of the local and the global. Russia–Latin America relations exemplify the impact that emergent complex cooperative networks are having on international order and organization because present engagement is explicitly premised on establishing long-term networks of cooperation to ground interstate relations.

DOI: 10.1057/9781137308139

Introduction

Abstract: *This part introduces the research problem and attendant research questions to be explored in greater detail. The section also introduces a general discussion of Complex Adaptive Systems (CAS) and discusses the potential contributions of this approach to the analysis of Complex Cooperative Networks (CCNs), the central theoretical contribution of this study for analyzing world politics in general and contemporary Russo-Latin American relations in particular.*

Astrada, Marvin L. and Martín, Félix E. *Russia and Latin America: From Nation-State to Society of States*. New York: Palgrave Macmillan, 2013. DOI: 10.1057/9781137308139.

Permanent peace and security are ostensible aims of all states in the international system. As representations of an enlightened international societal order, however, they remain elusive states of affairs. In a system of states primarily premised on fundamental realist assumptions such as the natural primacy of social conflict, the centrality of the nation-state as the fundamental political actor, and the absence of international authority or anarchy in international politics, it is difficult, if not impossible, to conceive of international peace and security evolving from any other source but power, force, and fear. In other words, according to the realist tradition in international politics, international peace and security are transitory conditions that may develop only by states remaining vigilant and ready to counterbalance by force potential adversaries in the international state system. Admittedly, since the inception of organized political units—especially in light of the need for such units to engage in basic, minimal contacts with one another—because no political unit exists in complete isolation—international peace, security, and order have proven to be volatile and intermittent states of affairs. History reveals that war has been a constant in interstate affairs, and the institutionalization of permanent international peace and security has been generally an unfulfilled project.

Throughout history, small-scale wars as the Athenian siege against the Melians as well as large-scale international wars such as the Great War of 1914–1918 and World War II in 1939–1946 have plagued attempts to promote and maintain the institutionalization of peace and security in international politics. This has effectively rendered international peace a temporary absence of war.[1] The threat or presence of international mass violence in either the Peloponnesian city-state system or the Westphalia nation-state system has led to the postulation that interstate peace and security are indicative of an impermanent absence of war.[2] Such reasoning dates back to Plato's time, as noted in *The Laws*, when he discusses the logic of laws that make it mandatory for citizen-subjects to engage in a militarized political program. In *The Laws*, Plato articulates that war, *not* peace, is the normal and most permanent state of affairs among politically organized units in interstate affairs. Accordingly, since the inception of interstate relations, this presupposition became the mainstay of its conceptualization and understanding. As Plato lucidly writes

> The aim of our institutions is easily intelligible to anyone... All these regulations have been made with a view to war, and the legislator appears

DOI: 10.1057/9781137308139

to me to have looked to this in all his arrangements.... He seems to me to have thought the world foolish in not understanding that *all are always at war with one another*; and if in war there ought to be common meals and certain persons regularly appointed under others to protect an army, they should be continued in peace. *For what man in general term peace would be said by him to be only a name; in reality every city [political unit] is in a natural state of war with every other, not indeed proclaimed by heralds, but everlasting.* And if you look closely, you will find that this was the intention of the Cretan legislator; all institutions, private as well as public, were arranged by him with a view to war; in giving them he was under the impression that no possessions or institutions are of any value to him who is defeated in battle; for all the good things of the conquered pass into the hands of the conquerors... *a well governed state ought to be so ordered as to conquer all other states in war.*[3]

Plato's reasoning, echoed by Thomas Hobbes, John Locke, Immanuel Kant, and many other political philosophers and contemporary international relations scholars over two millennia,[4] has been a foundational premise of the Realist conceptualization and understanding of international relations before and after the 1648 Peace of Westphalia (which gave rise to the notion of the modern system of sovereign states). Political Realism has, in turn, been the traditional, theoretical fundament of interactions between political units. The enduring ideas of Plato, Thucydides, Sun Tzu, Kautilya, Niccolò Machiavelli, Thomas Hobbes, Immanuel Kant, John Locke, Jean Bodin, Jean Jacques Rousseau, Carl M. von Clausewitz, and many modern thinkers such as E. H. Carr, Hans Morgenthau, George F. Kennan, Henry Kissinger, Kenneth Waltz, John Mearsheimer, and Herman Kahn, among others, have grounded explanations for the basic features of interstate relations. That is, power-politics, state-centrism, international anarchy, the self-help system, rationality, instrumental morality, and balance-of-power define international politics and order.

As articulated in Realist thought, the notion of the modern system of states is based on practical and material factors rather than on ethical, idealistic, and/or ideational objectives. In the twenty-first century, however, processes of globalization, including, but not limited to, political, economic, and social interfacing and integration of states via international organizations, have ushered in significant and variegated changes that promote alternative modes of attachment, interaction, and solidarity. Side-by-side with the mainstays of Realist thought there

DOI: 10.1057/9781137308139

has emerged the virtual, financial, cultural, and political interfacing via complex global networking. Networks have emerged—"globalist" technologies such as the internet have emerged, creating unprecedented layers of complexity and interconnectivity within interstate relations. At the broadest level, this work examines the question of whether these complex networks pose serious challenges to the rules of formation, structure, and dynamics of the system of sovereign states. The term "globalist" is part of the larger phenomenon of "globalism," which has been defined variously as an ideological, political, sociocultural, and/or economic program that actively promotes and develops the potential for states to transcend Realist limitations, such as international anarchy, on interstate relations. Realist explanatory frameworks view interstate relations as being constituted by self-interest, conflict of interests, lack of transparency, distrust, uncertainty, lack of shared ethical values and norms, and the absence of planned processes of coordination. Such a basis for interstate relations renders cooperation between states very difficult, mainly functional as opposed to contentiously integrationist, and wholly state-dependent rather than independently based on multiple and viable sources of international organization. Generally speaking, globalism, as Thomas L. Friedman writes, involves the

> inexorable integration of markets, transportation systems, and communication systems to a degree never witnessed before—in a way that is enabling corporations, countries, and individuals to reach around the world farther, faster, deeper, and cheaper than ever before, and in a way that is enabling the world to reach into corporations, countries, and individuals farther, faster, deeper, and cheaper than ever before.[5]

Globalism finds substantive expression in the form of international organizations that establish the basis for complex cooperative networks that link states in a meaningful way beyond mere, specified functionality. From a globalist perspective, international organizations promote "alternative modes of attachment"[6] when considering the objects and dynamics of interstate relations. "Alternative modes of attachment" can be viewed as fodder for the establishment of a "societal" framework of interstate relations, whereby complexity and cooperation complement and/or transcend the Realist limitations on the objects and dynamics of state interaction. Indeed, there is ongoing debate among scholars as to whether the "changes" inaugurated by globalization, on a descriptive, prescriptive, and substantive level, have, in fact, substantively impacted the system of states, and if so, to what degree.

DOI: 10.1057/9781137308139

The foregone discussion contextualizes the aim of this work. Our purpose is to examine the question of whether the emergence of complex networks has established viable, alternative bases of engagement for states when dealing with one another. In light of the integrative nature of the sundry strategic, cultural, technological, political, and financial developments that have appeared on the world stage, e.g., the internet, Facebook, Twitter, suppression conventions for effectuating international criminal law, and economic dispute resolution via the WTO, has the emergence of what can be termed hyper-complexity, deep networks, and exponential cooperation introduced a "societal" basis for interstate relations? Have processes of globalization resulted in the instauration of an embryonic society (as opposed to a mere system) of states?[7] In turn, does this international society build upon an atomistic system of states while concomitantly challenging its core strategic paradigm, anarchical ordering principles, and restrictions? To examine and flesh out these questions this work utilizes the notion of complex cooperative networks (hereinafter CCN) derived, in turn, from the *Complex Adaptive Systems* approach (hereinafter CAS).[8]

Under a CAS perspective, CCN may provide an alternative means for establishing a complex societal basis for state interaction. This is the case because a CAS perspective provides an effective means of mapping out and accounting for the systemic changes that have been taking place in world politics since 1945, when the international order underwent a seismic restructuring in form and substance. For example, the Nuremberg Tribunal introduced and formalized a radically new concept of universal justice into international relations in the late 1940s. For the first time in history, individuals, acting in a state capacity or on behalf of the state, were prosecuted for newly inaugurated "war crimes" and "crimes against humanity." In this sense, the Nuremberg Tribunal laid the groundwork for the viability of universal human rights as a condition to be factored into state policy and accountability to the international community. Further, nation-states have created CCN, like the League of Nations and United Nations, as a means of effectuating and giving substance to international legal norms and political frameworks to promote and effectuate concepts such as universal human rights which are alien and, admittedly, hostile to system-of-states-based ordering precepts such as absolute state sovereignty.

Analytically speaking, an international society can be likened to a complex social contract, wherein states surrender degrees of sovereignty for effectuating CCNs that then facilitate better cooperative basis

DOI: 10.1057/9781137308139

for state-to-state engagement and relations. Within a CAS perspective, individual CCNs exercise degrees of influence and autonomy that challenge the hierarchical order of the state-centric paradigm, which places states and not non-state actors at the apex of international order. International society, as a viable alternative to Realist explanatory schemata for interstate relations, has been emerging since 1945. This is the case because of the inauguration of various international organizational instrumentalities like the World Bank, the International Monetary Fund, and the World Trade Organization, among a host of others, which—*realpolitik* interests not withstanding—fostered an integrative notion of international order. For the past half century or so, international organization and global order has developed on the ideal that states can engage one another on a more complex basis other than *realpolitik*. Within itself, the post–World War II global order contained the seeds of fostering complex interaction and long-term relations premised on "universal" norms (e.g., universal condemnation of genocide) and principles of voluntary cooperation rather than on purely material and self-interested principles and considerations. The United Nations is a prominent example, as well as The League of Nations, the Nuremberg Tribunal, and the WTO, of states fashioning voluntary cooperative mechanisms (i.e., CCNs) to interface, integrate, cooperate, and promote incentives for interstate relations premised more on international societal and communal considerations than on purely strategic, materialistic, or *realpolitik* objectives.

A society of states, as a conceptual paradigm, though built upon an interstate system framework, has been ushering in a variety of transformative structural factors in international order since the end of World War II, particularly since the Bretton Woods System inaugurated the present world economic order. The instauration of the World Bank, for instance, was part of a larger globalist, i.e., integrative agenda layered onto the states system during Bretton Woods negotiations. The results of the Bretton Woods conference, whether or not inadvertent, did more than merely reenergize and rehabilitate a war-torn Europe while concomitantly enhancing and expanding American power. Also, it was the articulation and embodiment of a vehicle for the dissemination of a globalist or systemic economic agenda promoting the expansion of market economic practices and principles on the world stage. In other words, it was sort of an ideational program that took the form of an ideological and paradigmatic framework for international order, with

DOI: 10.1057/9781137308139

corresponding empirical and institutional practices and representations in the form of an integrated, networked, and universal economy. In short, the processes of economic globalization were and continue to be ideological globalism in action. In this sense, globalization, as material processes based on cooperative and integrative relations between states in the form of suppression conventions, treaties, agreements, and trade relations, is complemented by globalism, the ideational component of the ideological dissemination of liberal economic concepts, norms, and principles. As James Petras and Henry Veltmeyer note, globalization

> refers to the widening and deepening of international flows of trade, capital, labor, technology and information within a single integrated global market…it identifies a complex of changes produced by the dynamics of…development as well as the diffusion of values and cultural practices associated with this development. As a prescription, "globalization" involves the liberalization of national and global markets in the belief that free flows of trade, capital, labor, and information will produce the best outcome for growth and human welfare.[9]

The contemporary global order and its organization are, thus, characterized by diffusion of power from the exclusive purview of the state to the inclusion of myriad non-state actors as well. Non-state actors have increasingly assumed a critical role in international organization by creating and promoting complex social and institutional networks to facilitate and manage state-to-state relations. The International Court of Justice (ICJ), the Law of the Seas Convention, and the Anti-Hijacking and Terrorism Convention are examples of CCNs that foster cooperation on a complex societal level, and reflect greater degrees of voluntary acceptance of and adherence to international norms of cooperation, justice, and integration. Notions of international order based on a pure system of states perspective would relegate CCNs to the periphery of international relations and organization; CCNs would be viewed as nothing more than contrivances of states to be used selectively and dispensed with at will. Explaining the integration, influence, functions, and membership of newly empowered networks of non-state actors in world politics, Jonathan McClory notes that, these "may comprise a diverse set of actors, including states, civil society groups, NGOs, multilateral organizations and even individuals…[that] may form to tackle complex, trans-national collective action problems like Climate Change, environmental degradation, or take up single issues like banning landmines."[10]

DOI: 10.1057/9781137308139

In light of the posited questions and above discussion, this work will conduct a select analysis of Russo-Latin American relations to flesh out and examine how CCNs have impacted the changing and evolving nature of international relations by facilitating the instauration of a society of states. CCNs play a fundamental role in the development of a societal basis for interstate relations because they function within, complement, and manifest the application of soft power in international organization. The complex, integrative, networked, and adaptive nature of CCNs effectively provides states with a viable means of conducting their affairs without recourse to or minimizing the use of hard power. As this work will show, Russia–Latin America relations exemplify the highly effective and efficient use of CCNs as an expression of soft power in interstate relations and international organization.

Notes

1   See Félix E. Martín, *Militarist Peace in South America: Conditions for War and Peace* (New York: Palgrave Macmillan, 2006), pp. 9–17.
2   Ibid., p. 9.
3   Plato, "The Laws, Book 1 Selections," *The Human Condition: Philosophical Issues War & Peace: The Philosophy Resource Center*, accessed on 3 March 2010 <http://www.radicalacademy.com/hcwpfilehome5a.htm>. Emphasis added.
4   See, Immanuel Kant, *Perpetual Peace: A Philosophical Essay*, trans. and intro. by M. Campbell Smith, preface by L. Latta (London: George Allen and Unwin, 1917), pp. 117–119. Kant affirms that, "A state of peace among men who live side by side is not the natural state (*status naturalis*), which is rather to be described as a state of war: that is to say, although there is not perhaps always actual open hostility, yet there is a constant threatening that an outbreak may occur. Thus, the state of peace must be established. For the mere peaceful relations, and unless this guarantee is given by every individual to his neighbor—which can only be done in a state of society regulated by law—one man is at liberty to challenge another and treat him as an enemy." Thomas Hobbes asserts that "[out of civil states, there is always war of everyone against everyone. Hereby it is manifest that during the time men live without a common power to keep them all in awe; they are in that condition which is called war; and such a war, as is of every man, against every man.... so the nature of war, consisteth not in actual fighting; but in the known disposition thereto, during all the time there is no assurance to the contrary. All other time is PEACE." Thomas Hobbes, *The Leviathan*, ed. Michael Oakesshott, (New York: Collier, 1962), p. 100.

DOI: 10.1057/9781137308139

5   Thomas L. Friedman, *Longitudes and Attitudes*, 2002, accessed on 4 January 2011, <http://www.thomaslfriedman.com/longitudes prologue.htm>.

6   Michael J. Shapiro, *Methods and Nations: Cultural Governance and the Indigenous Subject* (New York: Routledge, 2004), p. xi.

7   See Hedley Bull, *The Anarchical Society: A Study of Order in World Politics*, 2nd ed., forward Stanley Hoffmann (New York: Columbia University Press, 1995). According to Bull, states form a system when they have a sufficient degree of interaction, and where each state is impacted by other states' decisions, interests, policies, etc. States behave, to some degree, as parts of a systemic whole. A system of states can exist without being a society of states. A society of states emerges when states, "conscious of certain common interests and common values, form a society in the sense that they conceive themselves to be bound by a common set of rules in their relations with one another, and share in the working of common institutions." For a more elaborate definition of system and its characteristics, see Robert Jervis, *System Effects: Complexity in Political and Social Life*, (Princeton, New Jersey: Princeton University Press, 1997), pp. 5–27.

8   See J. Poncela et al., "Complex Cooperative Networks from Evolutionary Preferential Attachment," *Plos One*, Vol., 3, No. 6, pp. e2449 (2008), accessed on 23 December 2012 available online <http://www.plosone.org/article/info:doi/10.1371/journal.pone.0002449>.

9   James Petras and Henry Veltmeyer, "Globalization or Imperialism?" in Daniel Egan and Levon A. Chorbajian (eds), *Power: A Critical Reader* (Upper Saddle River: Pearson, 2005), p. 184. See, also, UNRISD *States of Disarray: The Social Effects of Globalization* (Geneva: UNRISD, 1995); and World Commission on Culture and Development (WCCD), *Our Creative Diversity* (Paris: UNESCO 1995).

10   Jonathan McClory, *The New Persuaders II*, Institute for Government, 2011, accessed on 26 July 2012 <http://www.instituteforgovernment.org.uk/sites/default/files/publications/The%20New%20PersuadersII_0.pdf>

DOI: 10.1057/9781137308139

1

Mapping Complex Cooperative Networks: Society in International Organization

Abstract: *This chapter lays out the theoretical framework used to analyze interstate relations, exploring alternative international mechanisms and principles of order and organization. The emergence of soft-power complex cooperative networks in interstate relations has facilitated various changes in how states interact within the context of an international system rooted in realpolitik. Complex cooperative networks present challenges and opportunities for states within the traditional system-of-states order. Such networks, e.g., the United Nations, provide alternative bases for state engagement based on more complex interaction because of the integrative effect that networks have on the fabric of international order and organization. Russia–Latin America relations are employed to empirically explore the impact of emergent complex cooperative networks.*

Astrada, Marvin L. and Martín, Félix E. *Russia and Latin America: From Nation-State to Society of States*. New York: Palgrave Macmillan, 2013. DOI: 10.1057/9781137308139.

DOI: 10.1057/9781137308139

A society of states is premised on "soft power" notions and ordering principles. These include, among others, reputation, prestige, social capital, universal notions of justice, and international networks involving trade, finance conventions, diplomacy, and negotiations. In unison, these substantively impact, expand, and reconfigure the perceptions and behavior of international actors. According to *Monocle Magazine*, soft power refers to

> the ability of a state to achieve a desired outcome through the leveraging of legitimacy, or better still, attraction. Soft power eschews the traditional foreign policy implements of carrot and stick, relying instead upon...reputation and benevolent disposition...soft power opens up a third front, supplementing the traditional mediums of international relations: governments and [state foreign policy makers].[1]

Soft power, unlike "hard power"—viz., military capability, is distinct because states' influence in international affairs is relational and fluid in nature; concepts, perceptions, and interpretation take place in an evolving systemic context, giving rise to complex adaptive behavior among the constituent components of the international system. Jonathan McClory notes that soft power is

> the ability of a state to influence the actions of another through persuasion or attraction, rather than coercion...power can be wielded in three ways: threat of force (stick), inducement of payments (carrot) or shaping the preferences of others. Soft power eschews the traditional foreign policy implements of carrot and stick, relying instead on the attractiveness of a nation's institutions, culture, politics and foreign policy, to shape the preferences of others.[2]

Accordingly, soft power has played a cardinal role in the instauration of a society of states via Complex Cooperative Networks (CCN).

In light of the role CCNs have come to assume in international organization, do societal notions of international order rooted in soft power create viable alternative spaces for a conceptual framework(s) and set(s) of empirical practices that challenge the intelligibility of international order under a state system paradigm? Is it the case that the state, which under a system-of-states paradigm is "politically incontestable,"[3] is becoming less significant in an emergent society of states? Some commentators claim that there has been a continuous and relative decline of the nation-state in relation to other international institutional and

DOI: 10.1057/9781137308139

non-state actors in world politics and international organization. In fact, Petras and Veltmeyer affirm that,

> [n]otwithstanding considerable evidence of the state's continued promi-
> nence and agency within the global development process, *it is just as clear*
> *that under the present widespread structural and political conditions, the powers*
> *of the nation-state have been significantly eroded*, giving way to the influence
> of international institutions.[4]

This work does not support the above line of thought. Rather, CCNs as expressions of international society and soft power are viewed as functioning in tandem with Realist notions of international order. CCNs have not made the state irrelevant or impotent; rather, they have enabled the state to grow, adapt, and expand. States are complex entities that function within a system. Within the systemic context, states are provided with various means to better stabilize international affairs by creating integrative and cooperative bases for interaction. The dissemination of universal values such as human rights and economic development and prosperity have enabled the state to adapt to a shrinking world; integrative political, economic, and social institutions, in tandem with linking technologies such as the internet have provided the state with the tools it needs to remain viable, responsive, growing, adaptive, resilient and relevant in an ever-changing, networked, and interactive world.

Society, complexity, networks and Russia–Latin America relations

To explore the impact of CCNs on international order/organization, Russo-Latin American relations are used to test and illustrate the utility of the CAS approach and assess the contention that the instauration of a society of states in various international settings is taking place in the present global order. Relations between selected Latin American countries and Russia may provide a vivid example of how a society of sates is evolving via cooperative, interactive networks, especially through global economic processes, involving production, trade, finance, and investment.

In the case of Russia and Latin America, relations have undergone a profound reorganization and have been repurposed since the disintegration of the USSR and the end of the Cold War. Previous relations were

DOI: 10.1057/9781137308139

premised on Russia balancing US power, making incursions into the US's sphere of influence or "backyard," and supporting states on an ideological basis. Present relations are complex and variegated, based on a variety of levels—in particular trade, investment, and finance—that transcend the previous myopic basis for relations. For example, Sistema Económico Latinoamericano y del Caribe (hereinafter SELA) notes that the volume of trade between Russia and Latin American in the twenty-first century has almost tripled; relations have expanded in other economic areas that resemble society-forming attributes. Buttressing the argument of this work, the SELA study reports that

> [Russia] has expanded cooperation in...the energy sector, mining, physical infrastructure and telecommunications and...military technology [and] on projects in the areas of oil prospecting and extraction, construction of hydroelectric power stations, space exploration, and the use of nuclear energy for peaceful purposes...The strengthening of the economic, trade and financial links between Russia and the countries of the region [are] based on a legal framework of agreements which is permanently renewed with new agreements among governments and companies of both parties.[5]

Russia–Latin America relations, on a regional level, provide an interesting case study because global trade in particular has assumed the properties of a transformative mechanism vis-à-vis societal notions of international organizations that transcend traditional state-centric ordering principles. This work contends that trade, as a form of soft power and complex adaptation, has provided a basis for developing complex, systemic "institutional mechanisms," i.e., CCNs, for further enhancing and developing economic relations between Russian and Latin America. For example, while trade negotiations usually involve contacts between persons and organizations at the highest levels of government, negotiations entail various other actors that further expand the degree of integrative interaction. CCNs such as the OAS, MERCUSOR, ALBA, and the National Committee for Economic Cooperation with the Latin American countries are expanding and laying the basis for a sophisticated and complex basis for relations. The SELA study, also, reports that

> the actions that have been developed to foster the relations between Russia and the countries of Latin America...are very particular for each country depending on the hierarchy, priority, and level of existing relations, and in addition, the type of negotiations and the methods devised for their direction and control.[6]

DOI: 10.1057/9781137308139

Present Russia–Latin America relations exemplify how states—developing and developed—are coping with the transformative shifts ushered in by the diffusion of power as the result of economic globalization.

CCNs which have created alternative possibilities for state behavior and relations are producers and products of the massive changes that have been setting the stage for the rise of networks, for complex state-to-state engagement and relations. Emergent CCNs are being utilized by Russia and its Latin American partners to realize strategic as well as a host of non-strategic interests and goals. It is instructive to note here a central point emphasized by the SELA study which is directly related to the argument of this work and to the notion of reciprocity among multiple institutional entities in a variety of issue-areas:

> The legal and contractual basis for reciprocal cooperation is being con-solidated with new inter-governmental agreements among economic agents and scientific institutes of both parties. The intensification of the political dialogue and high-level visits has played an important role in improving this legal basis... [relations] go beyond merely economic aspects, as they also deal with political, scientific, technological, humanitarian and even military issues.[7]

A systems approach to international society and organization

How do CCNs emerge, and what enables them to act effectively so that states have an alternative basis from which to engage in interstate relations? CCNs are premised on the notion of adaptive, evolving systems comprising state relations as opposed to fixed systems of knowledge and understanding. A Complex Adaptive Systems (CAS) is comprised of layers of networked, interactive systems of knowledge that inform, complement, and produce opportunities and possibilities for the emergence of CCNs. Methodologically, General Systems Theory (GST) is based on the notion of "systems inquiry." As Bela Banathy notes, generally speaking the concept of "system" refers to a configuration of individual parts connected and joined together by a web of complex and interdependent relationships; in a specific sense, system is defined as a "family" of relationships among individual parts acting as a whole, i.e., as "elements in standing relationship."[8] Robert Jervis refines the

DOI: 10.1057/9781137308139

notion, holding that a system involves "(a) a set of units or elements [that] is interconnected so that changes in some elements or their relations produce changes in other parts of the system, and (b) the entire system exhibits properties and behaviors that are different from those of the parts."[9]

In general, a system, individually and collectively, is composed of regularly interacting parts which give rise to systemic activities. A set of concepts, whether empirical, metaphysical, or philosophical, work in tandem within an interdependent set of organizational relationships.[10] GST, like international relations, is multifaceted and thus engenders multidisciplinary approaches for conceptualization, theorization, analysis, and explanation.[11] International relations and organization are comprised of complex, interactive, interdependent, and interconnected systems. In the case of global security and economy, GST's systemic conception of international relations, in contrast to "a mechanistic one characterized by a [formal] description of structure and parts,"[12] is based on the interactivity of a variety of variables (parts) to produce a system of interaction, interconnectivity, and engagement beyond hard power principles of international organization. The difference between a "collection" and a "system" (whose parts are comprised of multiple sub-systems), as explained by Bertalanffy, is "that in a collection the parts remain individually unchanged whether they are isolated or together...whereas in a system the parts necessarily become changed by their mutual association; hence, their whole becomes more than just the sum of the parts."[13] Focusing on structure, while indispensable to explaining international relations, is too narrow in that it posits a closed system with various "laws" that apply across the board. A CAS approach is "a way of thinking having the proportions of a world view...as opposed to singular principles or parts of a structure."[14]

While international relations may seem to take place in a closed system, as is the case in Kenneth Waltz's seminal work, *Theory of International Politics*,[15] it is a rather complex and open system that is in constant flux, comprised of various constituent epistemic and material components. Networks that produce and are produced by systemic interconnectivity are interactive, adaptive, dynamic, and multidimensional and broaden possibilities for alternative forms of engagement. States in the international system are thus viewed as sets of "organized actions which are maintained constantly by exchanges [and interaction] in the environment."[16] As Ahmed, Elgazzar, and Hegazi note in their overview

DOI: 10.1057/9781137308139

of Complex Adaptive Systems, two interrelated definitions are necessary to grasp these organic entities

(1) a complex adaptive system consists of inhomogeneous, interacting adaptive agents. Adaptive means capable of learning, transforming, and adapting.

(2) "[an] emergent attribute of a CAS is a property of the system as a whole which does not exist at the individual elements (units or agents) level...to understand a complex system one has to study the system as a whole."[17]

Further, these authors claim that a "systemic approach is against the standard reductionist one, which tries to decompose any system to its constituents and hopes that by understanding the elements one can understand the whole system."[18]

CAS posits that individual agents (parts or units) are the collective base elements of the system that interact, and then adapt in response to interactions, thus allowing for innovation and maximization of the potential for the individual parts to realize and work cooperatively toward fulfilling common self-interests and goals. The utility of the CAS approach for our purpose is its value to elucidate the complex and decentralized nature of CCNs in the present international order.[19]

It is important to underscore here, as John H. Holland makes clear, that a CAS

> has no single governing equation, or rule, that controls the system. Instead, it has many distributed, interacting parts, with little or nothing in the way of a central control. Each of the parts is governed by its own rules. Each of these rules may participate in influencing an outcome, and each may influence the actions of other parts. The resulting rule-based structure becomes grist for the evolutionary procedures that enable the system to adapt to its surroundings.[20]

A CAS can, thus, be conceptualized, as Howard Bloom notes, as a

> learning machine, one made up of semi-independent modules which work together to solve a problem. Some complex adaptive systems, like rain forests, are biological. Others, like human economies, are social....Both apply an algorithm—a working rule—[that sets the foundational context(s) for conception, perception, and realms of possibility and actuality].[21]

CAS are, accordingly, "complex" because they are diverse, composed of multiple interconnected components and are adaptive because

DOI: 10.1057/9781137308139

a system has the capacity to change, to enable the individual parts to "learn" alternative modes of interaction. Adaptation enables states to go with the ebb and flow of the changing dynamics that are characteristic of a networked, integrated system. Arguably, states can, therefore, "learn" new ways of perceiving, strategizing, maximizing their utility with new tools embodied in CCNs. Cooperation within a competitive yet highly integrated and networked system in a systemic context is an outgrowth of states "learning" to ameliorate international anarchy and its consequences.[22] According to Kevin Dooley, a CAS behaves and evolves according to three key principles:

1 Order is emergent as opposed to predetermined;
2 The system's history is irreversible;
3 The system's future is often unpredictable—the basic building blocks of the CAS are agents that interact with each other and their environment.[23]

In the case of international relations, interaction, while subject to change and "evolution," also retains degrees of consistency, regularities, and continuity, enabling a network-based system of governance to emerge despite the prevalence of systemic anarchy in the international context. Thus, interactive and interdependent relations between agents that are situated in macroscopic collections of interacting units are endowed with the capacity to evolve and adapt to a fluid environment.[24] In the case of international relations, a systematic set of concepts and practices (tangible and intangible) are utilized by international actors who work in tandem within an interdependent set of organizational relationships to conceive, interpret, articulate, and implement globalist notions of international order.[25]

Russia–Latin American relations provide a working case study of this phenomenon. For example, Russia's Minister of Foreign Affairs met with Cuba's Minister of Foreign Affairs on July 11, 2012. Both parties emphasized the importance of growing and strengthening bilateral cooperative frameworks in the form of treaties and agreements, and both countries confirmed similar perspectives regarding supremacy of international law and the central role of the United Nations in the conflict negotiation and settlement.[26] Such pronouncements and agreements, as will be discussed in subsequent chapters, typical of Russia's "new" *modus operandi* within the region, are providing the foundation for numerous CCNs to emerge between Russia and Latin America.

DOI: 10.1057/9781137308139

The emergence of a society of states within a system of states is indicative of a displacement of a multiagent framework (hereinafter MAF) to a CAS. A MAF signifies

> a loosely coupled network of problem-solver entities that work together to find answers to problems that are beyond the individual capabilities or knowledge of each entity... each agent has incomplete capabilities to solve a problem; there is no global system control; data is decentralized; computation [rational cost-benefit analysis] is asynchronous.[27]

What distinguishes a CAS from a purely MAF structure or entity is the focus on the macroscopic properties that transcend traditional ordering principles of state-to-state interaction (e.g., anarchy and balance of power)—properties such as complexity. A MAF describes the traditional system of states because it is as a system composed of states that are sovereign and that interact within an anarchical systemic context.

In a CAS, the agents as well as the system are complex adaptive actors that distinguish and learn the quantitative and qualitative differences between optimal and sub-optimal outcomes. Complexity and adaptation procure self-similarity on a systemic level. Self-similarity involves the notion that a self-similar object is approximately similar to the system in which it is emplaced, and has similar properties as one or more of the parts that constitute the system—coastlines, for example, are statistically self-similar in that parts of them show the same statistical properties at many scales.[28] Self-similarity also applies to states; i.e., individuated sovereignty, e.g., finds expression in systemic anarchy—the two are based on the property of sovereignty, and each feeds into the other. Self-similarity, within the context of an emergent society of states, enables CCNs to create intimate and inextricable networked ties of connectivity based on systemic and systematic engagement that creates venues of cooperation that then attenuate the effects of an anarchical global context based on material power and balancing of power to attain a temporal cessation of hostilities and contextual volatility.

Agents, CCNs, and international organizations

The fact that international relations take place in a systemic context does not necessarily imply that there are no degrees of agency available to the constituent components of the system. "States are rarely found

DOI: 10.1057/9781137308139

in complete isolation from one another. Most inhabit relatively stable systems of other independent states which impinge on their behavior."[29] Within the system-of-states context, while there are overarching ordering principles that delimit certain structural parameters, viz., anarchy, material considerations, security, and balance of power, states (parts or units) have degrees of agency as to how best to effectuate power augmentation, attain self-interests, and obtain security. Sovereignty, in theory, provides the bases for diversity in state perceptions and behavior. Hence, political, social, cultural, and economic organization takes a variety of forms in the international system.

Consensus over the exact meaning of the term "agent" is not necessarily obvious. Alexander Wendt contends that agency takes place within a bounded universe in international relations, and is subject to the following constrictions: (1) physical security, viz., territorial integrity and plenary control of territory; (2) ontological security, which entails the need for relatively stable expectations about the world; and (3) sociation, which entails the need for minimal levels of contact and interaction between parts, given that states do not exist in a vacuum.[30] These principles shape perceptions of power and self-interest, and the tactics and methodologies by which states fulfill their primary directives, i.e., survival, development, and perpetuation. Agency complicates the ordering principles because it involves the notion of "adaptive learning." That is, parts are subject to socialization processes based on "sociation." Learning enables the parts to comport their behavior to the rules of engagement defined by the systemic material and ideational ordering principles in place. "Structures have effects not reducible to agents.... The structure of any social system will contain three elements: material conditions, interests, and ideas."[31] Agency is situated in the realm of ideas and to a lesser extent the interpretation of interests.

Within the context of a system or society of states, agency is both individuated (state level) and collective (systemic level). Within a society of states, agency produces or rather is the precursor for the emergence of a collective intelligence, reflected in CCNs, which, in turn, can be considered the products of technologies and strategies of cooperation. Indeed, such technologies and strategies of cooperation have complex adaptive potential because of the fact that globalism is rooted in three of the most basic and "universal" structures that under-gird all social order, i.e., some form of commerce (trade).[32] In the case of states, they can be viewed as "intelligent agents" that produce but are also products

DOI: 10.1057/9781137308139

of the systemic international context. Agents share common attributes, viz., degrees of adaptation, autonomy, individual and collective interests, and collective goal-orientation. In addition to these attributes, the agents engage in collaborative, cooperative behavior, which facilitates a collective intelligence to deal with the various challenges that arise from the anarchic global context. CCNs, as agents, possess a capacity for enabling states to create aggregated knowledge bases via cooperative technologies and strategies based on collective intelligence and learning capacities.

CCNs are impacting interstate relations. A global economic and socio-political and cultural mesh, network, is not a new phenomenon—indeed, international affairs have been "World Wide Webbed and Internetted [*sic.*] since Rome began to import silks from China in roughly 200BCE."[33] Yet, the prominence and growing reliance on and independent efficacy of CCNs have ushered in different modalities of power-diffusion and mass communication that differ from previous manifestations. In particular, CCNs possess what Flores-Mendez describes as an "inferential capability," i.e., an ability to act upon and effectuate specific goal-orientated tasks based on collective or systemic "intelligence." This ability, combined with mobility—whereas states are fixed in territorial space—reactivity, i.e., the ability to selectively process and act upon data, and "temporal continuity," i.e., the persistence of identity, lends CCNs degrees of influence and power independent from states in a society of states paradigm.[34]

CCNs can be classified as having "weak" or "strong" degrees of agency, depending on the level of integration, subject matter, level of professional expertise, financial resources, and issue-area or topic relevance to the more powerful state actors or a majority of the international community. Whether weak or strong, agents in a CAS are interactive, complex entities whose contacts and engagement go beyond basic, minimal contacts that being in a system implies. The systemic context of anarchy establishes a shared space and bounded space of engagement. This shared space provides the basis for shared cooperative rules of engagement that, in turn, help shape the rules of formation vis-à-vis the character and content of relations, interactions, which creates ties based on shared knowledge and communication.

In the case of international organization and the variegated CCNs it is interesting to note that under a CAS, CCN-as-agents have created or rather laid the foundations for high degrees of global integration absent within a purely system-of-states paradigm. CCNs, as alternative venues for the resolution of conflict and the basis for cooperative, networked

DOI: 10.1057/9781137308139

interaction, provide alternative infrastructures upon which to base state-to-state interaction. As Flores-Mendez notes, "[i]nfrastructures provide the regulations that agents follow to communicate and to understand each other, thereby enabling knowledge sharing."[35] Infrastructures in the form of networks—e.g., legal instruments such as treaties, government officials, and global regulators—are viable in that they provide the following for integrating states on the world stage: i.e., a networked ontology, which allow agents (CCNs and states) to agree on the meaning of fundamental ordering concepts such as the Universal Declaration of Human Rights; communication protocols, which describe a common, collective discourse, language, for communication; communication infrastructures, which specify legitimate, authoritative channels for communication; and interaction protocols, which describe the means and methods for proper state-to-state interaction.[36]

The argument is that CCNs provide infrastructures that reflect and reify, what Edward Fischer has termed, "cultural logics."[37] Cultural logics are "predispositions, grounded in history and tradition but also responding to immediate circumstances, of how to look at (and act in) the world in certain ways. Cultural logics provide the basis for culturally logical improvisations."[38] CCNs, as agents of governance, present states with the potential for new social structures, new cultural logics, more centralized and cooperative approaches and solutions to problems, new international norms, and alternative means to obtain objectives and define purpose and identity.[39]

Conclusion

The above discussion of CCNs within a CAS context provides a comprehensive, conceptual, and principled framework for explaining the role of complexity, networks, cooperation, integration, and society in interstate relations. Subsequent chapters will further refine the concepts and principles discussed above and, then, examine Russia–Latin America relations to flesh out how CCNs are impacting thought and practice vis-à-vis statecraft and state-to-state engagement on the world stage.

In the case of Russia–Latin America relations the respective countries seem to have concluded that CCNs provide a basis for more constructive relations. While it is certainly the case that states retain sovereignty, and that *realpolitik* considerations remain a priority for states, it is,

DOI: 10.1057/9781137308139

nevertheless, the case that "social orders are rooted in a multiplicity of times, trajectories, and rationalities that, although particular and sometimes local, cannot be conceptualized outside a world that is globalized."[40] Within the context of Russia–Latin American relations, the respective countries appear to have "learned" that alternative forms of engagement, rooted in societal, cooperative, soft-power notions of trade, investment, legal, and cultural exchange, seem to complement traditional forms of engagement. Since 2000 until present, relations have expanded to include a variety of non-state actors, institutions, and practices based on deep networks premised on economic exchange, scientific, and technological sharing and cooperation, and financial investment and cultural exchange between the respective countries.[41]

Notes

1   Monocle Magazine, *A Briefing on Global Affairs, Business, Culture & Design*, "Soft Power Survey," Vol. 4, No. 39, pp. 41–50 (December 2010/January 2011). The article can be found at http://stage.monocle.com/magazine/issues/39/the-new-soft-sell/. Also see Joseph S. Nye, Jr., *Bound to Lead: The Changing Nature of American Power* (Cambridge: Harvard University Press, 1990); and Joseph S. Nye, Jr., *Soft Power: The Means To Success in World Politics* (New York: Public Affairs, 2004).

2   Jonathan McClory, *The New Persuaders: An International Ranking of Soft Power*, Institute for Government December 2010, accessed on 21 July 2012 <http://www.scribd.com/doc/47790532/THE-NEW-PERSUADERS-An-international-ranking-of-soft-power>. See Joseph Nye, "Public Diplomacy and Soft Power," *Annals of the American Academy of Political and Social Science*, Vol. 616, March (2008), pp. 94–109.

3   Michael J. Shapiro, *Methods and Nations: Cultural Governance and the Indigenous Subject* (New York: Routledge, 2004), p. 7.

4   James Petras and Henry Veltmeyer, "Globalization or Imperialism?" In *Power: A Critical Reader* Daniel Egan and Levon A. Chorbajian (eds) (Upper Saddle River: Pearson, 2005), p. 189. Emphasis added

5   Sistema Económico Latinoamericano y del Caribe, "Economic Relations Between the Russian Federation and Latin America and the Caribbean: Current Situation and Prospects," Caracas, Venezuela, July 2009, accessed on 21 July 2012 <http://www.sela.org/attach/258/EDOCS/SRed/2009/07/T023600003569-0-Economic_relations__Russian_Federation_and_LAC.pdf>.

6   Ibid.

DOI: 10.1057/9781137308139

7   Ibid.

8   Bela Banathy, "A Taste of Systemics," *The Primer Project: A Special Integration Group (SIG) of the International Society for the Systems Sciences (ISSS) (originally SGSR, Society for General Systems Research) and the International Institute for Systemic Inquiry and Integration: The First International Electronic Seminar on Wholeness*, December 1, 1996 to December 31, 1997, 25 February 2008 http://www. newciv.org/ISSS_Primer/asemo4bb.html. Also see Ludwig von Bertalanffy, *General System Theory: Foundations, Development, Applications* (New York: George Braziller, 1968); Ludwig von Bertalanffy, *A Systems View of Man*, ed. Paul A. LaViolette (Boulder: Westview Press, 1981).

9   Robert Jervis, *System Effects: Complexity in Political and Social Life* (Princeton: Princeton University Press, 1997), p.6.

10 See Ervin Laszlo, *The Systems View of the World: A Holistic Vision for Our Time Advances in Systems Theory, Complexity, and the Human Sciences* (Cresskill: Hampton Press, 1996); Ervin Laszlo, *The Systems View of the World: The Natural Philosophy of the New Developments in the Sciences* (New York: George Braziller, 1972).

11 For examples of the application of systems analysis/inquiry in international relations, see R. J. Crampton, *The Hollou Détente: Anglo-German Relations in the Balkans 1911–1914* (Atlantic Highlands: Humanities Press, 1980); James Rosenau, *Turbulence in World Politics* (Princeton: Princeton University Press, 1990); Inis Claude, *Power and International Relations* (New York: Random House, 1962); Morton Kaplan, *System and Process in International Politics* (New York: Wiley, 1957); Jack Snyder and Robert Jervis, (eds), *Coping With Complexity in the International System* (Boulder: Westview Press, 1993).

12 Richard B. Gray, ed., *International Security Systems: Concepts & Models of World Order* (Itasca: F. E. Peacock, 1969), p. 2.

13 Bertalanffy, *Systems View* ix.

14 Charles McClelland, "General Systems Theory in International Relations," *International Security Systems: Concepts & Models of World Order*, ed. Richard B. Gray (Itasca: F. E. Peacock, 1969), p. 21.

15 Kenneth Waltz, *Theory of International Politics* (Reading, Mass.: Addison-Wesley, 1979).

16 McClelland, "General Systems," p. 22.

17 E. Ahmed, A. S. Elgazzar, and A. S. Hegazi, "An Overview of Complex Adaptive Systems," *Mansoura J. Math*, 28 June 2005, 10 May 2010, p. 6.

18 Ibid., p. 7.

19 See Kevin J. Dooley, "A Nominal Definition of Complex Adaptive Systems,'" *The Chaos Network*, Vol. 8, pp. 2–3 (1996); Kevin J. Dooley, "A Complex Adaptive Systems Model of Organization Change, *Nonlinear Dynamics, Psychology, and Life Sciences*, Vol. 1, pp. 69–97 (January 1997).

DOI: 10.1057/9781137308139

20   John H. Holland, "Complex Adaptive Systems," *Daedalus*, Vol. 121, No. 1 (Winter 1992) pp. 17–30, particularly, see pp. 21–22.

21   Howard Bloom, *Global Brain: The Evolution of Mass Mind from the Big Bang to the 21st Century* (John Wiley: New York, 2001), p. 9.

22   See M. Mitchell Waldrop, *Complexity: The Emerging Science at the Edge of Order & Chaos* (New York: Penguin, 1992); Jason Brown Lee, "Complex Adaptive Systems," Technical Report: Complex Intelligent Systems Laboratory, Centre for Information Technology Research, Faculty of Information Communication Technology, Swinburne University of Technology, Melbourne, March 2007, accessed on 21 July 2012 <http://www.scribd.com/doc/22947963/Complex-Adaptive-Systems >.

23   Kevin Dooley, "A Nominal Definition of Complex Adaptive Systems," pp. 2–3.

24   "Complexity in the Social Science Glossary: A Research Training Project of the European Commission," no date, accessed on 2 April 2010 <http://www.irit.fr/COSI/glossary/fulllist.php?letter=M>.

25   See Ervin Laszlo, *The Systems View of the World: A Holistic Vision for Our Time Advances in Systems Theory, Complexity, and the Human Sciences*; Ervin Laszlo, *The Systems View of the World: The Natural Philosophy of the New Developments in the Sciences*; Marvin L. Astrada, *American Power after 911* (New York: Palgrave Macmillan, 2010).

26   Ministry of Foreign Affairs of the Russian Federation, "On the meeting of the Minister of Foreign Affairs of Russia S.V. Lavrov and the Minister of Foreign Affairs of Cuba B. Rodríguez," July 11, 2012, accessed on 27 July 2012 <http://www.mid.ru/bdomp/brp_4.nsf/e78a48070f128a7b43256999005bcbb3/1ddbd4d4188cc78e44257a390024ced7!OpenDocument>.

27   Roberto A. Flores-Mendez, "Towards the Standardization of Multi-Agent System Architectures: An Overview," ACM Crossroads, Special Issue on Intelligent Agents, Association for Computer Machinery, Issue 5.4, pp. 18–24 (Summer 1999) available online (pp.1–12, p. 4, 23) accessed on 1 December 2012 <http://gicl.cs.drexel.edu/people/regli/Classes/KBA/Readings/flores-acm-crossroads.pdf >. See also N.R. Jennings, K. Sycara and M. Wooldridge, "A Roadmap of Agent Research and Development," in *Autonomous Agents and Multi-Agent Systems Journal*, N.R. Jennings, K. Sycara and M. Georgeff (eds) (Kluwer Academic Publishers: Boston, 1998) Vol. 1, No.1, pp. 7–38; E.H. Durfee, *et al.*, "Trends in Cooperative Distributed Problem Solving," in *IEEE Transactions on Knowledge and Data Engineering*, (March 1989), KDE-Vol. 1, No. 1, pp. 63–83.

28   See Benoît Mandelbrot, 1967, How Long Is the Coast of Britain? Statistical Self-Similarity and Fractional Dimension. Science, New Series, Vol. 156, No. 3775. (May 5, 1967), pp. 636–638.

DOI: 10.1057/9781137308139

29 Alexander Wendt, *Social Theory of International Politics* (Cambridge: Cambridge University Press, 1999), p. 10.

30 Ibid., pp. 131–32.

31 Ibid., p. 139.

32 Howard Bloom, *The Genius of the Beast: A Radical Revision of Capitalism* (Amherst: Prometheus, 2010), p. 22.

33 Ibid., p. 47.

34 Flores-Mendez, "Towards the Standardization of Multi-Agent System Architectures: An Overview," p. 2.

35 Ibid., p. 4.

36 Ibid. See Neil MacCormick, "Beyond the Sovereign State," *Modern Law Review,* Vol. 56 (1993); Moisés Naím, "Five Wars of Globalization," *Foreign Policy* (January/February 2003), pp. 29–36; Renaud Dehousse, "Regulation by Networks in the EU: The Role of European Agencies," Journal of European Public Policy, Vol. 4 (1997), pp. 246–261; Robert O. Keohane, "Governance in a Partially Globalized World," Presidential Address, Annual Meeting of the APSA, 2000, *American Political Science Review,* Vol. 95 (March 2001), p.1; Robert O. Keohane and Joseph S. Nye, Jr., "Trans-Governmental Relations and International Organizations," *World Politics* Vol. 27 (1974), p. 39; and Robert O. Keohane and Joseph S. Nye, Jr., *Power and Interdependence: World Politics in Transition* (Boston: Little, Brown, 1977).

37 See Edward F. Fischer, *Cultural Logics and Global Economies: Maya Identity in Thought and Practice* (Austin: University of Texas Press, 2001), pp. 1–10.

38 Edward F. Fischer, "Summary Analysis: Guatemala Strategic Culture," June 1, 2010 (unpublished manuscript). Summary prepared for Florida International University.

39 Bloom *Genius,* p. 117.

40 Achille Mbembe, *On the Postcolony* (Berkley: University of California Press, 2001) p. 9.

41 See e.g., Sistema, "Final"; Eduardo José González, "Russia and Latin American Strengthen Judicial Cooperation," Radiohc, March 30, 2012, accessed on 28 July 2012 <http://www.radiohc.cu/ing/news/cuba/6210-russia-and-latin-american-strengthen-judicial-cooperation.html>.

DOI: 10.1057/9781137308139

2

Globalization, International Organization and the Rise of a Society of States

Abstract: *This chapter explores the emergence of complex cooperative networks as a tool of Statecraft. Such networks have substantively impacted the dynamics of international order and organization. Soft-power based networks, products of globalization processes, have integrated states to an unprecedented degree. The networking of states' interests, purpose, and wellbeing, e.g., rule of law, economic growth and prosperity, sustainable development, regulation and administration of global interaction between states, and the dissemination of universalistic notions such as universal human rights, is characterized by the fusion of the local and the global. In the case of Russia–Latin America relations, present engagement is premised on establishing long-term networks of cooperation in the realms of energy, trade, investment, and cultural exchange, among others.*

Astrada, Marvin L. and Martín, Félix E. *Russia and Latin America: From Nation-State to Society of States.* New York: Palgrave Macmillan, 2013. DOI: 10.1057/9781137308139.

DOI: 10.1057/9781137308139

Society involves basic, minimal contacts such as trade, with interaction taking place among politically organized units that are "self-conscious and self-regulating entities."[1] A system of states is formed when at least two states have engaged in minimal contacts that impact the respective states' perceptions and conduct, with each state acting in concert with one another in a systemic context.[2] Society develops when states go beyond basic minimal contacts, and engage in complex behavior that has the effect of deeply networking and integrating the parts. Each social entity, that is, system and society, involves and is based upon the nature and degree of interaction among politically organized units, with interaction based upon overarching ordering principles that guide relations and behavior within an order. Under a society paradigm of global order, "the more states are in contact with one another and agree to the same principles, the more they homogenize."[3]

The modern system of states has been committed to sovereignty, to preserving the integrity of state supremacy within designated geopolitical borders. Strategic calculations based on hard-power capacity, such as military technologies, have been the norm—the *élan vital* of state behavior since the founding of the modern states system in 1648. Sovereignty has been the preeminent value and ordering principle of international relations since the Peace of Westphalia.

> A commitment to sovereignty drove the internal and external dimensions of the modern state. Modern states adhered to absolutist notions of sovereignty in their internal policy choices and external policy and military objectives. The notion of sovereignty was intricately linked to the nation as a constitutive feature of the state, to welfare as the legitimating drive of the state, and to the balance of powers as a key component of the state's external strategy.[4]

In the post–World War II era of international relations and organization, defined by globalization, a structural transformative shift has taken place, and rigid adherence to notions of geopolitical borders and sovereignty are "gradually losing the central role they played in the modern era."[5] The emergence of Complex Cooperative Networks (CCNs) as a tool of statecraft has changed the dynamics and fulcrum of power on the world stage.

CCNs, in thought and practice, are the nodes that stitch together the super-network ushered in by globalization. As a phenomenon, the processes of globalization have a multifarious capacity due to the fact that

DOI: 10.1057/9781137308139

globalization has the potential to span the spectrum of possibility as far as being endowed with high degrees of political, ideological, cultural, economic, and social interpretations of international order. Since the institutionalization of the modern global political and economic order, the organization and management of world affairs have been premised on an intimately networked, integrated global liberal-economic system. Free-market economics and procedural democracy have been the structural organizing principles of the global politico-economic system established at the end of World War II. Although various definitions have been proffered, Helga Turku offers a useful characterization of globalization as

> the processes that constitute globalization involve the transformation of local states affairs (political, social, and economic) into global states of affairs. Increasingly, people around the world have integrated into a common society, based on a particular mode of global governance rooted in market economics and procedural democracy. In its most general definition, globalization involves the concentrated combination of social, economic, technological, socio-cultural, and political forces.[6]

Globalization is a product as well as producer of the complex super-network and sub-networks emerging on the world stage that are, in turn, fomenting a society of states. Society indicates a substantive transformation of international order that involves grafting societal notions onto the preexisting system of states. Globalization is the *élan vital* of the structural evolution that is taking place in the fabric of international order and affairs.

> In the political analysis of globalization, the term is often utilized in a "doctrinal sense" to refer to neo/liberal economic globalization, the dominant process that integrates domestic economies into an overarching international economy by way of global trade, foreign direct investment, uninhibited capital flows and labor migration, and the dissemination of science and technology...globalization also has a political dimension that works hand in hand with economic philosophy and policy.[7]

Globalization thus involves processes that span the realms of domestic, territorial notions of order and fuse the political, sociocultural, and the economic, encompassing ideational variables such as ideology and culture and material variables such as commerce and trade. Accordingly, globalization impacts the totality of international society and not just a specific aspect of it, that is, the economy.[8]

DOI: 10.1057/9781137308139

The super-networking of states' interests, purpose, and weal—which includes, but is not limited to, security (global and domestic), political stability, law, economic growth and prosperity, sustainable development, the necessity of regulation and administration of global interaction among and between state and non-state actors, and the dissemination of universalistic notions of global order such as the notion of human rights—is characterized by the fusion of the local and the global in the realms of politics, economy, and culture. In the case of Russia–Latin America relations, since 2008 Russia has maintained extensive and intensive contact and engagement with Latin America, with such engagement premised on establishing long-term networks of cooperation in the realms of security, energy, trade, investment, military affairs, military modernization and capacity building. After a tour of the region (2008), which included Venezuela, Cuba, Brazil, and Peru, the then-Russian President Medvedev declared:

> We have visited states that have never been visited either by Russian or Soviet leaders before. This means only one thing: no attention has been paid to these countries. In a sense, we are only just starting fully-fledged, full-format and, I hope, mutually beneficial contacts with the leaders of these states, and with the economies of these states, respectively. There is nothing to feel shy about; one should not fear competition here. One should bravely join in the fight.[9]

Although there are certainly strategic calculations involved in Russia's decision to expand its relations with Latin America, perhaps, an unintended consequence is that the means by which to obtain strategic interests inadvertently employs societal mechanisms by which to do so. The use of CCNs to obtain strategic interests is part of the complexity of state perception and behavior, and the indelible effect of employing CCNs is to legitimate and reify societal notions of international order. CCNs become more and more relevant, indispensable, to statecraft, and provide viable alternatives to the sometimes counterproductive use of resources and negative effects of employing force to obtain a state's goals. Continued use of trade, investment, cooperative ventures in all aspects of relations is part of the "learning" that takes place, learning that involves embracing values and conduct that are not in line with a purely states system view of international order. Indeed, universal notions of human rights, justice, and the renunciation of war as a tool of foreign policy (e.g., the Kellogg-Briand Act) are examples of the type of values that are

DOI: 10.1057/9781137308139

expressed in CCNs and that are taking hold in the present configuration of international order and interstate relations.

Although the states system remains intact to the extent that states remain the dominant actors on the world stage, and that anarchy remains endemic to structuring relations among global actors, international order in the post–World War II system, particularly since the collapse of the Soviet Union and the end of the Cold War in 1989 and how it subsequently played itself out in the early 1990s, is indicative of a complex adaptive systems (CAS) that has become intimately integrated, interactive, complex, and interdependent. The high degrees of "intensification of worldwide social relations, which link distant localities in such a way that local happenings are shaped by events occurring miles away, and vice versa,"[10] is comprised of a CAS that is itself the product of the innumerable fusions that are taking place among and between various CCN sub-systems. Full-blown complex interdependence has become a hallmark of the CAS that emerged after the 1970s, even though the processes that comprise globalization are not a novel phenomenon.

Already in 1848, Karl Marx observed that, in "place of the old local and national seclusion and self-sufficiency, we have intercourse in every direction, universal interdependence of nations."[11] The "universal interdependence of nations" has produced networks of deep connectivity—culture, politics, regulation, economy, human rights, security, law, and diplomacy, among other states of affairs—with CCNs as the basic nodes that link up and maintain vibrant circuit interaction. The complex and integrative nature and character of the post-Bretton Woods international order, as Helga Turku notes,

> has produced high degrees of sociopolitical and economic interactivity and interdependence via extensive [and intensive] networks of state-to-state engagements and arrangements. Mass media communications, capital flows, environmental issues, international institutional regulations and arrangements, international treaties and legal policies, and international military relations and security policies, render state-to-state interaction and interdependence inevitable.[12]

Two basic pillars of globalization, that is, trade and communication, have had a profound effect on the fabric of the states system, especially in the ongoing process whereby the domestic and the foreign are becoming less distinguishable. Systemic responses to five major ongoing developments that peaked during the twentieth century vis-à-vis an emergent

DOI: 10.1057/9781137308139

society—i.e., (1) an international system of trade and finance and the free movement of capital; (2) threats that do not observe national boundaries (for example, disease); (3) loss of state control over homogenous culture; (4) commodification of weapons of mass destruction; and (5) a global system of human rights which imposes legal rules on nation-states, whether or not they have ratified them—have profoundly altered the dynamics of international engagement and organization.[13] As Turku notes, governments and societies across the globe are adjusting to a world in which there is no longer a clear distinction between domestic and foreign affairs.[14]

> Networks, complex webs of intensive and extensive connectivity, have emerged, and the aggregate of these networks has given rise to a CCN. In the case of Russia-Latin America relations, military cooperation rooted in hard-power-based strategic interests is complemented by global trade. Military cooperation has been complemented by other forms of cooperation. Energy cooperation between Russia and Venezuela has been welcomed in the form of oil drilling, Cuba is working with Russia to explore oil drilling options and Argentina and Brazil have engaged Russia by signing agreements to develop nuclear power capacity.[15]

Since 2000, Russia–Latin America relations have been building upon CCNs between high-level government officials, judicial agencies, military and security partnerships, and private sector cooperative agreements in the realm of energy and cultural exchange. Such networks have exponentially increased the capacity to exchange information and coordinate activities to address different issues that transcend a reductionist approach to interstate relations limited to security and the projection of military power. Global economic ties have been established via intense and intertwined networks.

Networking has permeated every aspect of present international order and organization. Indeed, CCNs have adapted and evolved and have augmented their capacity to effect and affect global relations due to the emergence of a society of states within a system of states. CCNs—whether in the form of treaties, conventions, transnational corporations, trade partnerships, cultural exchanges, and/or humanitarian relief organizations—embody shared knowledge of legal, security, defense, political, culture, and development procedures, methods, and/or enforcement mechanisms. In the realms of law and politics, there have been global cooperative systems established between various

DOI: 10.1057/9781137308139

agencies and departments between different countries, e.g., INTERPOL. As Turku affirms

> Despite the diversity underlying the various types of networks function-ing in the international system, global networking involves common functions, such as: the expansion of regulatory reach that allows national governments to keep up with transnational corporations, non-state actors, and transnational criminal enterprises, and building trust and establishing relationships among participants that create incentives to creating working relationships.[16]

For example, in the case of Russia–Latin America relations, Russia has actively forged complex trade and strategic ties with important regional players that include Venezuela, Argentina, Bolivia, Brazil, Cuba, Ecuador and Peru. As reported by Walter Walle

> In November 2009, Ecuadorian President Rafael Correa signed far-reaching pacts with Moscow regarding cooperation on security and defense... with the help of Moscow [Ecuador] hopes to develop nuclear technology to meet a portion of its energy needs... [Also, in] April 2010, Bolivian President Evo Morales asked [for] a greater Russian presence in the Southern Hemisphere... Bolivia, like Venezuela and Ecuador, has also invested in Russian technology... Furthermore, Moscow approved a 100 million USD credit line for La Paz in order to purchase a variety of military equipment, such as helicopters to combat drug trafficking, and a new presidential aircraft.[17]

The establishment of cooperative networks has laid the foundation for a complex and adaptive systemic CCNs based on sustainable, intimate ties of long-term cooperation and the establishment of "universal" sources of knowledge and information, such as human rights, that become pro-ducers and products of a collective learning machine. It should be noted that, although the knowledge (and modalities) produced from the glo-balization super-network is the object of violent resistance, viz., political and religious-based terrorism, there are idealistic, "universal" principles that seem to transcend the particular ideological components of such globalized political programs such as the war on drugs, the war on terror, and efforts to effectuate economic modernization and development. It is not implausible to suggest that the deep networks that are proliferating may indeed have the potential to challenge the *realpolitik* notions that underlie international relations. Integral in this process has been the active role of CCNs such as the IMF, the World Bank, the United Nations,

DOI: 10.1057/9781137308139

the redefined role of the North Atlantic Treaty Organization (NATO) in Europe since the collapse of the Soviet Union, the WTO (World Trade Organization), the Red Cross, Amnesty International, the World Health Organization (WHO), and the Organization of American States (OAS), among others.[18] Such CCNs have assumed degrees of agency to impact the fabric of and operative bases for implementation of global policies, be they of an economic, security, developmental, cultural, social, health, or educational related nature.

Further, CCNs have assumed a profound role in the conduct of day-to-day affairs on the world stage. In the case of Russia–Latin America relations, CCNs have emerged in the realms of combating drug trafficking and strengthening judicial ties. To expand and enhance its strategic and economic interests, Russia has actively sought to cultivate relations with Latin America based on providing a plan that involves a non-militarized cooperative international effort rather than a myopic militarized effort. Also, as reported by Walle

> In contrast to American planned initiatives, like Plan Colombia or the Mérida Initiative for Mexico, aimed at combating the trafficking of drugs through military means, Russia has proposed the "Rainbow-3," a plan to manage the drug trade through development and job creation. [The plan] would raise the drug issue to a new level of international involvement, through the UN Security Council...focusing on 'the elimination of the social causes of drug production, such as unemployment and poverty'...[Russia would] provide special training and custom-tailored courses for the police forces of Central American countries at no cost to their governments.[19]

Also, as reported by *Pravda* in March 2012, Russia's efforts have been publically acknowledged and welcomed by Latin American leaders. Nicaraguan President Daniel Ortega, for instance, went on the record stating:

> "We are grateful for the efforts by the Russian government to strengthen the fight against drug trafficking in Nicaragua"...Minister of Foreign Affairs of El Salvador Hugo Martínez said that the initiative will be useful at the regional level. "Russia's experience in this field can help the fight against drug trafficking, not only in El Salvador but also within the Central American Integration System" (Belize, Guatemala, Honduras, Dominican Republic, Costa Rica, Nicaragua, Panama and El Salvador).[20]

Similarly, the observation made by Keck and Sikkink regarding advocacy networks is readily applicable to CCNs in the context of diverse and

DOI: 10.1057/9781137308139

expansive networks in Russia–Latin America relations, i.e., networks share

> the centrality of values or principled ideas, the belief that individuals can make a difference, the creative use of information, and the employment by non-governmental actors of sophisticated political strategies in targeting their campaigns... [CCNs] are bound together by shared values, a common discourse, and dense exchanges of information and services.[21]

Thus, CCNs are actively involved in attempting to shape international affairs based on a societal as opposed to a purely state-centric notion of international affairs. With the advent of CCNs, interstate relations have become more open, fluid, and inclusive of societal notions of international order. Society is able to thrive in an anarchical context because of sociality. As A.N. Whitehead observes, symbolic signification possesses a protean capacity; symbols can have different meanings for different people under different circumstances.[22]

A society is able to transcend the system of state's limited conceptual discourse and applied methodologies governing options, interests, and interaction on the world stage. In the case of Russia–Latin America relations, many of the countries involved promote an alternative to the present world order, i.e., countries seek to implement a "multipolar" form of international organization. Through the Sistema Económico Latinoamericano y del Caribe, a CCN comprised of Russia and different Latin American countries that promotes trade and economic integration, member states believe that international organization in the twenty-first century should be "independent of the unipolar imperial trend [that] emerged... after the fall of the Berlin wall."[23] Multipolarity is

> a notion shared by Venezuela and many of the geopolitically independent Latin American and Caribbean governments. In the case of Venezuela, this notion of multi-polarity is strongly rooted in the Bolivarian conception of external relations, which is one of the fundamental pillars of Venezuela's foreign policy... of achieving a universal balance... to build a pluripolar, multipolar or polycentric world.[24]

In practice, CCNs exert degrees of influence and provide possible avenues for engagement that are not available under a system of states. Albeit in a different context and analysis, Karns and Mingst's observations help define how CCNs' activities shape interstate relations, because CCNs

DOI: 10.1057/9781137308139

try to set the terms of international and domestic debate, to influence international and state-level policy outcomes, and to alter the behavior of states...and other interested parties.... Network analysis encompasses both international and domestic actors and processes.[25]

Because CCNs possess this potential, they pose challenges to a state-system conception of international order. Challenges, however, does not imply incompatibility or irreconcilable differences. Indeed, it seems that states have "learned" how to harness the power of CCNs for obtaining goals. The societal principles that animate and underlie CCNs, namely, cooperation, do have the political, social, and ideological implications for the conduct of international affairs under a states system perspective. To illustrate this point, former Assistant Secretary of State and UN Ambassador John Bolton notes

> [I]t is precisely the detachment from governments that makes international civil society so troubling, at least for democracies...Indeed...the civil society idea actually suggests a "corporativist" approach to international decision-making that is dramatically troubling for democratic theory because it posits "interests" (whether NGOs or businesses) as legitimate actors along with popularly elected governments.[26]

CCNs do represent an emergent shift from "government" to "governance," effectuating a "significant erosion of the boundaries separating what lies inside a government and its administration and what lies outside them."[27] In this context, the (so-called) "erosion" of state power and its monopoly over the ideational and material mechanisms by which to define the character and content of international affairs is under assault from society. Yet, it appears that states have used CCNs and the societal principles they embody to their benefit, having alternative tools of statecraft that are parallel to and that actually complement rather than erode state power. Granted, the dynamics of power have been impacted, and states do find themselves heavily dependent on certain CCNs, namely, those that facilitate and effectuate global commerce, nonetheless states are poised to better comprehend and react to the networked nature of the present international order. Howard Bloom's observations about networks shed light on why CCNs can enhance state power, and why states like Russia can adapt and "learn": that is, because CCNs are part of

> an immense and purposeful union of interacting parts—a nearly infinite agglomeration of networks within networks and machines within

DOI: 10.1057/9781137308139

machines...Networks [that] stretch and alter to achieve new capabilities... [that] mesh as modules in yet grander webs of being.[28]

The contentiousness between society and states system principles (i.e., those that revolve around the power to legitimately posit the identity, roles, and rules of formation and interaction of actors in international) does not stand as an insurmountable obstacle to functioning concomitantly in the conduct of interstate relations.

CCNs: adaptive learning and international order

CCNs flourish, replicate, and become efficient and effective because they enable states to perceive, process, "learn," store, manipulate, interpret, re-interpret, and disseminate information in ways that fundamentally expand viable alternative options for constructive engagement. CCNs retain high degrees of efficaciousness and viability because of the network-based origins of their power. While states are bounded by territory and limited options for engagement, CCNs provide a diverse array of options that transcend territoriality, e.g., using the internet as a political, social, and economic tool. As Karns and Mingst note, one can describe the transnational activity of CCNs as

> networks and coalitions [that] create multi-level linkages between different organizations that each retain their separate organizational character and memberships, but through their linkages enhance power, information sharing, and reach.[29]

A system of states, on its own, propagates a less dynamic and limited meta-language of knowledge, truth, power, and control—rooted in modernist concepts and practices, such as physical territory and military power as the *sine qua non* of international affairs—that, standing alone, are losing efficaciousness without the complement of governance and society to sustain a state's power to legitimately define the interests, purlieus, etc., of international affairs and order. CCNs present an alternative meta-language to international order, and this is "legitimate" because of the fact that in all configurations of order there is, as Jean Francois Lyotard and Jean-Loup Thebaud observe, "no meta-language...that [can] ground political and ethical decisions that will be taken as the basis of its statements. There is no meta-language; there are only genres of language, genres of discourse."[30]

DOI: 10.1057/9781137308139

The CAS–CCN genre of language, discourse, is more suited to a hyper-networked and interlinked global order due, in part, to the ability of CCNs to effectively and efficiently identify, process, learn, store, generate, and research information, as well as disseminate and integrate information and technology due to a "memetic" capacity. A meme is "a habit, a technique...a sense of things, which easily flips"[31] from individuated part to another part of a larger system of interaction. CCNs are memetic in that they are able to generate data of international affairs in a way that non-networked entities, viz., states, are not due to the adaptive, fluid and networked nature of CCNs. This enables CCNs to develop a legitimate and viable power-base that possesses degrees of independence from the traditional mainstay of power, authority, and legitimacy in the international system, that is, the state. Memetic learning involves not only empirical based but also ideational transmission of knowledge across space and time. The meme itself is conceived as an "immaterial replicator which duplicates itself in the virtual soup of minds."[32] While the meme utilizes mimesis, it also produces substantive change in the "collective mind" of the international community, affecting perception and conduct on the level of values and operative frameworks for comprehending and responding to eventuation through policy and action. Universal notions of international order—universal human rights, universal justice, condemning genocide, norms of democratic political organization and free-market economics being the "end of history"—are the manifestation of the memetic capacity of CCNs to thrive in a context that excludes or minimizes principles and actors of international organization not premised on the primacy of the territorial state. The global integration of the states system has resulted in an exponential "knowledge explosion and the evolution of data webs with a whole new style,"[33] wherein learning and the transmission of knowledge occurs via cabling actors internally (via values, norms, and principles, for instance) as well as externally (via trading partnerships and security alliances, for instance). Memetic learning enables an array of international actors to pool information in a way that surpass the ability of the isolated state; and given that the state relies upon CCNs to effectuate global learning processes in a world defined by intimate, integrated, complex interdependence, CCNs seem to have the potential to garner a significant degree of power when it comes to impacting, contouring, and/or controlling international agenda(s).

DOI: 10.1057/9781137308139

CCNs create a viable basis for a "common" culture on the global level. When goals, interests, and conceptual, perceptual, and empirical methodologies become "common," the commonality of society provides an alternative to the commonality of the individuated state residing in a system of anarchy. Ludwig Wittgenstein is correct in noting that the limit of one's language constitutes the limits of one's world.[34] Discourse helps form the parameters and contours of global "reality." This has the effect of producing a base of knowledge that centers around particular power-concepts, systemically and systematically perpetuating and reifying a "correct" discourse of right and authority. The power-concepts embodied in discourse can be analogized to analytic nets that are placed over a particular object. That is, in the words of Wittgenstein

> different nets correspond to different systems for describing the world. Mechanics determines one form of description of the world by saying that all propositions used in the description of the world must be obtained in a given way from a given set of propositions—the axiom of mechanics. It thus supplies the bricks for building the edifice of [knowledge], and it says, "any building that you want to erect, whatever it may be, must somehow be constructed with these bricks, and these alone."[35]

In the case of Russia–Latin America relations, commonality of interests are erected on the network-based building blocks of economic development, trade, and investment as opposed to a purely *realpolitik* basis (that is, the discourse of a purely states system perspective). For example, and as partial evidence of this development, as reported by SELA, joint economic development projects in several areas have been undertaken by Russia and countries in the region. These are

> primarily related to exploration and extraction of hydrocarbons, electricity, mining, mechanical engineering and transport.... Several co-investment projects have been implemented... with Venezuela, Colombia, Brazil, Bolivia and Cuba, among others... Russian technical cooperation with several countries in the region was also resumed in the energy field. Russia participated in hydroelectric projects in Argentina, Brazil, Bolivia, Chile, Colombia and Mexico. As regards civilian nuclear energy, Russia maintains cooperation projects with Venezuela, Mexico, Argentina, Chile, Brazil and Cuba.[36]

Thus, if we view CCNs as being products as well as producers of a CAS, and CCNs as mechanisms of trans-mutative learning based on the effective and efficient use of mimetic notions of hierarchy, information

DOI: 10.1057/9781137308139

pooling, collaborative information processing, and imitation, then CCNs exercise degrees of influence vis-à-vis subject matter and issues that were once the sole prerogative of the state, i.e., defining, impacting, and facilitating international organization and order.[37] The argument here is similar to the CAS dynamic discussed by Roy J. Eidelson when he argues that

> Through a dynamical, continuously unfolding process, individual units within the system actively (but imperfectly) gather information from neighboring units and from the external environment. This information is subjected to local internal rules, and responses are formulated; these responses then work their way through the web of interconnected components. Within the CAS, competition operates to maintain or strengthen certain properties while constraining or eliminating others. Entirely new properties can also emerge spontaneously and unexpectedly.[38]

Thus, conceptually, a CCN, as a derivative of CAS, can be viewed as a social learning machine that has the capacity to infiltrate state-centric principles of anarchy, balance of power, the sovereign state, and the like, that claim a monopoly over defining international organization. The use and primacy of force in international relations certainly does not become effaced from global affairs, but various complexities and externalities based on intimately and inextricably integrated networks now come into play in addition to the use of force and the accumulation and projection of material power. Machiavelli's notion that gold does not win wars, for example, is perhaps an antiquated concept due to a globalized economy, whereas the power of CCNs to significantly impact and contour international affairs via soft power has become a viable proposition.[39]

Accordingly, a CAS perspective provides a framework for alternative interpretations and postulations of global order. The lexis of a networked system enables "[n]ew ways of seeing" that "lead to new ways of being. Words are our lenses, our looking glasses, and our tools. They can refashion more than the way we see...words can reshape reality."[40] One does not have to subscribe to idealism in order to observe that it is possible for concepts and ideas to substantively and substantially impact thought and practice. New ways of perceiving international actors, state relations, non-state actors in global governance, the use of soft-power strategies and technologies, and the very fabric of order on the world stage are part of "new" rules of formation propagated by a society-of-states ethos.[41]

DOI: 10.1057/9781137308139

Components of CCNs: dynamics and implications

The components of the collective networked learning machine that a CAS posits are *Conformity Enforcers*; *Diversity Generators*; *Inner Judges*; *Resource Shifters*; and *Intergroup Competition*.[42] *Conformity Enforcers* function to ensure that a collectivity possess a variety of similarities, for example, cultural, ethnic, racial, civic, political, ideological, economic, and moral, in order to provide a basis for a shared worldview, collective identity—one that is unified, homogenized, via layers of conformity-inducing networks. The Nuremberg Tribunal is perhaps the most striking example of the phenomenon of complex adaptive learning, because it instituted a regime whereby states began to "learn" what types of conduct qualify as offences to all humankind—i.e., crimes against humanity.[43] The states system's lexis and enforcement methods, based on geopolitical and geostrategic premises of thought and action based on a coherent and cohesive national (state) culture and discourse, differs from a society of state's ordering premises of soft-power lexis and rules of formation. *Diversity Generators* enable a collectivity to retain a trajectory, spatio-temporal viability by way of spawning innovation via variety. In an almost dialectical process, networked modalities (viz., political, sociocultural, and economic) compete for prominence, dominance, and the more effective and efficient modalities that meet the basic as well as more complex needs and wants of a collectivity are retained while others are discarded. Innovation in technology, economics, natural science, and political organization—in sum, in every aspect of a collectivity's being—enables notions of progress and evolution to manifest. *Inner Judges* are ideational constraints based on macro-level notions of morality, proper norms, correct values, that serve to designate what is of value to the collectivity (for example, free trade), what behaviors are to be rewarded (for example, procedural democracy) or punished (for example, terrorism, ethnic cleansing, genocide). *Resource Shifters* range from social to economic systems, and involve mechanisms that facilitate knowledge and learning networks by distributing resources, utility, etc., to more efficient and effective modalities of governance. Lastly, *Intergroup Competition* involves collective learning systems having to integrate, cooperate, and interact in such a way as to produce an overarching and more efficacious template for governance through a competitive process.

The numerous, diverse, and multifarious interacting units that constitute a CAS are typically arranged in a hierarchical structure.[44] H.A.

DOI: 10.1057/9781137308139

Simon has described the arrangement of a CAS as "sets of boxes nesting within sets of boxes" through several repetitions.[45] Similarly, John Holland argues that new hierarchical levels are created whenever individual units, or agents, are aggregated. Aggregation thus initiates a basic foundation for the power and efficacy of CCNs, that is, a global context defined by hierarchy, networks, and interdependence. John Holland views aggregation as one of the most basic elements of a CAS. Eidelson observes that

> In turn, each aggregate can connect with other aggregates to form meta-agents, which can then combine to form meta-meta-agents, and so on. In this way, the aggregation of business firms forms an economy [and a basis for constructive engagement with independent rules of formation]. Some aggregations, such as human society, involve an especially complex network structure because each agent may belong within a number of different boxes at the same hierarchical level[46]

Collective learning machines are thus complex and hierarchically networked via "memes," that is, the "habits, new ways of doing things, and other commanding intangibles which migrate from [system] to [system]—[that] are key to the next jump up in networking."[47] Within the context of globalization and CCNs, memes can be classified into two distinct categories or types, that is, external and internal. External memes involve mechanisms associated with or that affect deep patterns of behavior, interaction, and perception that comprise the structural level of engagement between international actors (for example, anarchy, rationality, territory, balance of power) and internal memes are those that transfer socialization processes to the individual social units that comprise a system.[48]

Hierarchy does not imply that there is a centralized executive agent that is responsible for directing the totality of a CAS. The influence exercised by CCNs stem from and can be conceptualized within a framework, as explained by Eidelson, of the notion of

> communities of practice—informal networks of people [and organized and/or institutionalized collectivities] ranging from chemists in competing pharmaceutical firms to the foreign affairs personnel of adversarial countries to gangs in schools and prisons—[that] develop interaction structures that do not rely on central controls.[49]

This characteristic is part of the complexity that underlies the CAS, and why conformity and diversity are not mutually exclusive states of affairs.

DOI: 10.1057/9781137308139

Each works in tandem with the other for the more efficient conduct of the systemic whole. This is the case because collectivity is the basic unit of organization for systemic learning "machines," systemic learning processes. Although conformity breeds a sense of holistic identity and a homogenized basis for identification and identification processes, diversity is crucial because of the need to maintain high degrees of flexibility when it comes to an adaptive capacity vis-à-vis externalities that directly impact optimal efficiency and effectiveness, for example, anarchy in international relations and the role of CCNs in better facilitating international order. The notion of memes as "units of imitation"[50] has an ideational basis for their efficacy in the realm of international relations and international organization.

Challenging the system of states paradigm

The fact that a practice or rubric of power is or has been the case does not assure the necessity or exclusive justification of using a particular conceptual framework to comprehend international relations and organization.[51] Thus, a system of states paradigm and its rules of formation have no intrinsic superiority, but rather depend on contemporary processes and states of affairs.[52] Indeed, in an evolving and hypernetworked world, society seems to be the more effective and efficient means by which to effectuate governance to manage and administer international affairs. In the case of a societal CAS, memetic behavior enables changes to occur in orientation, perception, and behavior because of the fact that international "reality" or rather actuality is malleable, despite some basic structural principles that cannot be thought out of existence, such as anarchy if one retains the states system as the basic ordering template for the conduct of international affairs. The following observations help account for the role of memes in facilitating learning and knowledge distribution among subjects. That is, the reality of international order can be conceived as a "composite unity by a network of components that (i) through their interactions recursively regenerate the network of interactions that produced them, and (ii) realize the network as a unity in space [and time] in which the components exist by constituting and specifying the unity's boundaries."[53] International affairs are thus comprised of "a domain of interlocked (intercalated and mutually triggering) sequences of states, established

DOI: 10.1057/9781137308139

and determined through...interactions between structurally-plastic state-determined systems."[54] Basic, minimal contacts provide the basis for the evolution of more complex, intricate, and inextricable ties that transcend mere self-interest or minimal basic contact. Complex interdependence is a result of the grafting of network upon network, *ad infinitum*, resulting in a CAS steeped in societal notions of international order and organization.

Perceptions of international organization and actuality have changed, perhaps evolved, and this phenomena stems from the deep levels of interaction that transcends the most basic needs of states. "Individual perception untainted by others' influence does not exist. A central rule of large-scale organization goes like this: the greater the spryness of a massive enterprise, the more...communication [and interaction] it takes to support the teamwork of its parts."[55] Morality, legality, economy, politics, trade, ethnicity, race, culture—all of these states of affairs are joint and several, comprised of systemic networks that make cohesion possible, giving the international community its most important tool in the post-modern age, that is, society.[56] Brazil's "trade diplomacy" is an example of this phenomenon. Brazil has used its soft power resources to create mass CCNs in order to better link itself to trading partners. CCNs have resulted in Brazil being able to better harness its potential based on tangible (e.g., natural resources) and intangible (e.g., cultural) resources at its disposal. Brazil is presently "the world's fourth largest food exporter, and is the world's 16th largest oil producer. What Brazil has done differently than other Latin American countries is that it vehemently looked for new markets near and abroad."[57] Brazil has been able to develop its potential and further its wellbeing (which includes projection of power, values, and interest) via CCNs. Brazil recently signed a trade agreement with Israel,

> where it is expected to increase trade in the sectors of education, science, agriculture, and medicine, cementing Brazil's status as Israel's largest trading partner in Latin America," and it also signed investment agreements with Kenya and trade deals with China that includes a plan to "build a new steel plant in Brazil by the Chinese.[58]

Such networked systems, intimately and inextricably interlinked, align, confine, define, and synchronize the perceptions and actions of a collectivity and of a collective learning machine.[59] Thus, the discourse, the symbolic matrix, which a system-of-states paradigm uses to

DOI: 10.1057/9781137308139

conceptualize international affairs and posit international order, is being challenged by an emergent society. Territorial integrity, the "invention of [a nation-state cultural] tradition,"[60] of legitimate state-monopolization of violence, sovereignty, as the singular entity that defines, interprets, and conveys legitimacy to a homogenous, coherent national identity, and singular actor on the world stage—all of these states system notions, symbols, lexis, etc.—are faced with alternative interpretations. The "clarity" and logical inevitableness of a system of states paradigm, based on cultural homogeny and physical territorial boundaries, simply does not address the hyper-networked nature and complexity of a society of states.

CCNs are network-based tools of cohesion that can be utilized to effectuate society. Culture networks embodied in globalized political, social, and economic processes are the means by which CCNs, as products and producers of international society vis-à-vis a system of states ameliorate the deleterious effects resulting from multi-layered, state-centric interactions. CCNs, as the agents of a societal-networked international order, provide a measure of security and stability, ameliorating the destabilizing effects of anarchy. Social, political, and economic stability at the global level—elusive under a system of states paradigm or framework—becomes possible via the push and pull of sociality, conformity, diversity, while globalist notions of international order, such as human rights, becomes possible in a CAS comprised of CCNs as the facilitators of shared, collective knowledge and practice. "New" notions, concepts, and discourse of a globalist ilk—especially trade and investment that facilitates cooperative relationships between states—"up-grade the tools with which [we] think, the tools with which we go for insights in our daily lives. New symbols…give…new insights into…the nature of the [integrated] daily dealings between"[61] human beings and how societies engage in social organization and interaction on the global level. When considering Russia–Latin America relations, it is interesting to note that the

> parties have sought ways to go beyond the usual trade and financial relations and to establish more up-to-date and secure forms of cooperation, based on sectors with higher degrees of technological diversification and advancement. In this direction, they have been trying to foster cooperation in the energy sector, oil and gas extraction and transportation, machinery production, metallurgy, transport sector, aviation technology, nuclear energy for peaceful purposes and space exploration.[62]

DOI: 10.1057/9781137308139

Globalization involves concomitant centripetal and centrifugal dialectical processes; the ebb and flow of implementing universal, communitarian values, norms, interests, and the like, creates a need for networks, and this need has irrevocable consequences on the fabric of international order. CCNs constitute and provide "collective assemblages of enunciation"[63] on a global scale that articulate globalist as opposed to purely statist notions and interpretations of world order. The biologic principle of the Baldwin Effect, wherein "once an advantageous behavior has been embraced by a population, it will gradually reshape the genetic string of the species which has adopted it, resulting in biological rewiring,"[64] can be likened to the effect CCNs are having on the "DNA" of international order, viz., advantageous behaviors, practices, concepts, etc., of international society are altering and reshaping the initial system of states paradigm, gradually "rewiring" the political, economic, and social premises that ground international organization. The growth of what Howard Bloom terms "long-distance productivity teams"—the facilitators of global commerce, a "new" lexis of political, sociocultural and economic interaction, and free trade—are embodied in CCNs, government and non-governmental, the abstract as well as empirical nodes of interconnectivity that define a society of states.[65] "Knowledge in the form of an informational commodity, indispensable to productive power is already and will continue"[66] to play a major role in the generation of a societal paradigm and discourse of global order.

System and society: utterly antithetical paradigms?

Although Society poses challenges to a pure system of states paradigm, society is not necessarily antithetical to the system of states. Indeed, society has evolved from the states system, and retains a relationship with the state form of organization, viz., the state being the legitimate and authoritative repository of force, military power, on the world stage as long as sovereignty remains a cardinal value and premise of what are termed international relations. Sovereignty, the unfettered exercise of power within a state's geopolitical and legal borders, is somewhat at odds with governance because of the non-bounded nature of CCNs. Although the modern notion of sovereignty has variegated interpretations, one of the earliest and enduring definitions of the modern concept of state sovereignty in international relations is proffered by Jean Bodin (1530–1596).

DOI: 10.1057/9781137308139

For Bodin, "sovereignty is the most high, absolute, and perpetual power over the citizens and subjects in a Commonwealth, which the Latins call *Majestas*."[67] Furthermore, for Bodin,

> a sovereign prince is one who is exempt from obedience to the laws of his predecessors and more importantly, those issued by himself. Sovereignty rests in being above, beyond or exempted from the law.... Although it occupies a subordinate place in Bodin's theorization, it could be said that this exception from being subject to the law is the quintessential condition of sovereignty in so far as it is understood politically.[68]

A state exercising supreme power and authority over a geopolitical (cartographic) and legally defined and recognized (conceptual) territory, while possessing a monopoly over the use of force to compel obedience has, generally speaking, under-girded the notion of sovereignty. A state's strategic interests have produced and are a product of a system of states, and strategic thought and interests represent the rational, calculative, materialist, "objective" dimensions of state identity and power.[69] Military capacity, natural resources, geographic considerations, technological capacity, level of economic development, the projection of force and coercive power—these factors are inextricably linked with the perpetuation and augmentation of sovereign state power in a system of states. The US Department of Defense defines strategy as a "prudent idea or set of ideas for employing the instruments of national power in a synchronized and integrated fashion to achieve theater, national, and/or multinational objectives."[70] Security vis-à-vis strategic thinking is based in a militarized framework based on *realpolitik* considerations.[71] A strategic orientation toward security therefore involves:

> The overall relative power relationship of opponents that enables one nation or group of nations effectively to control the course of a military or political situation; the course of action accepted as the result of the estimate of the strategic situation. It is a statement of what is to be done in broad terms sufficiently flexible to permit its use in framing the military, diplomatic, economic, informational, and other measures which stem from it; the level of war at which a nation, often as a member of a group of nations, determines national or multinational...strategic security objectives and guidance, and develops and uses national resources to achieve these objectives. Activities at this level establish national and multinational military objectives; sequence initiatives; define limits and assess risks for the use of military and other instruments of national power; develop global plans or

DOI: 10.1057/9781137308139

theater war plans to achieve those objectives; and provide military forces and other capabilities in accordance with strategic plans.[72]

Thus, at the most basic level, the interest of the state in maintaining its territorial integrity has contoured conceptions, perceptions, and practice as to what constitutes international order. "Our idea of what belongs to the realm of reality is given for us in the concepts which we use."[73] Conflict, violence, and war—mainstays on the world stage—have, among several other reasons, been waged by the state to protect basic territorial integrity, which, concomitantly, is inextricably linked with sovereignty, which, in turn, is intimately associated with other key system of states ordering precepts, viz., balance of power, rational/strategic thought and interests, and military power as the *sine qua non* of state power.[74] Societal notions of order, premised on soft-power notions of order, such as the institutionalization of diplomacy, substantive cooperation, negotiation, and the instauration and connectivity of states via CCNs—which has resulted in complex, hyper-networks wiring state and non-state actors in ways that are unprecedented—are indeed antithetical to rigid and static notions of a territoriality and a singular, myopic focus on traditional notions of national security based in a systems of states prism of world order and the conduct and underlying currents and determinants of state-to-state relations and world affairs in general.

Ironically, although war and conquest are viewed as antithetical to societal notions of international order, each has been indispensable to procuring a society of states. Reciprocity has by no means been "the only human agglomerator [*sic.*] and information exchanger. There was also a data connector whose power to produce two-way information flows is filled with irony—war and conquest."[75] War, merger, and expansion through militarized conquest

> and subsequent consolidation has played a key role in the historic develop-
> ment of governance from spatially small units of authority—band, tribe,
> kingdom—to larger units. The anarchic strong nation system, thus, is the
> cumulative, historical by-product of political consolidation of innumerable
> smaller governance units and their integration.[76]

The violent processes of empire building, the brutal acquirement of territory and (slave) labor, and the like, laid the network foundations for global commerce—beyond the realm of pecuniary gain. Empire builders, for example, the Romans, Ottomans, and Greeks, "spread the tendrils pulling us together to this day. They broke the barriers separating mini-

DOI: 10.1057/9781137308139

groups by standardizing languages, writing systems, laws, trade, weights and measures, and by building roadways over which their troops could march and on which merchants, pilgrims, and the curious could follow."[77] Howard Bloom observes that conquest enabled memetic "learning" to take hold on a massive scale, thereby extending an information network. "India was a largely tribal culture until roughly the 6th century B.C., then the improvement of the plow led to agricultural surplus, a cash economy, and the rise of businessmen. A new breed of ruler discovered it could tax the commercial classes, use the money to build a professional army, and set out to subjugate every neighboring territory in sight. Expansion was built into the system."[78]

Conquest has thus been the engine that has integrated disparate units and sub-systems into a more comprehensive and cohesive network. Conquest, a fundamental principle of realist political thought, laid the basis for the global sociopolitical, cultural, and economic super-network that has integrated the states system in the present. Conquest "is the needle which stitched together virtually all of the 'great nations' which we know today—allowing such multi-tribal hodgepodges such as the State of Russia, Germany, Japan, the United Kingdom, and France to be the repositories for *ein Volk*, that is, one folk with a unique bloodline and history."[79] *E Pluribus Unum* is an accurate notion that captures the essence of the "tug of sociality" that underlies the integration of sub-systems of identity, culture, race, ethnicity, political systems, etc., into an overarching system referred to as a globalizing integrated and networked international system on a scale that has truly made strict application and adherence to some traditional ordering mechanisms archaic, e.g., territoriality as a primary basis for state power and the conduct of interstate relations.

In a society of states, CCNs thus become nodes of connectivity, flash-points, and synapses, for the proliferation of sophisticated, interactive, mutually dependent and supportive networks. In the modern states system, CCNs function as data connectors, based primarily on reciprocity, contracts (rules), economic and rational self-interest. Cultural, economic, political and other social linkages create a powerful basis for international order that complicates the states system. A CAS thus develops when the various sub-systems coalesce, are grafted onto each other, and mesh to form a super-network comprised of multifarious networks embodied in globalization. "A complex adaptive system is a 'nested hierarchy'—a net whose nodes are each part of a larger entity. Each of these larger 'super-organisms,' in turn, is a node in a far larger web. And each

DOI: 10.1057/9781137308139

is also a hypothesis."[80] Each sub-system embodied in specific cultures, political systems, ideologies, ethnic and/or civic identities, and the like, is a hypothesis as to a proper ordering of affairs. Through the interplay of and competition among variegated systems and sub-systems a mesh of interdependent decision-making centers, ideational transfer networks in the form of universal values, norms, political organization, and the like, merge and proliferate in a CAS context.

CCNs, as the nodes that facilitate the interconnectivity of various networks, give rise to a "functionally coherent and discrete cooperative assembly."[81] In the realm of economic and cooperative networks that bestride state and society notions of international organization, Russia and Latin America have attempted to build upon rational self-interest via governments utilizing cooperative networks to establish, enhance, and perpetuate economic development and wellbeing. These are indications of a growing sense of networked relations intended, as the SELA Report states, to

> diversify the external economic relations of Latin America and the Caribbean, strengthening relations of cooperation with the Russian Federation is of particular relevance. The frequent contacts at the highest political level between Russia and Latin American and Caribbean countries are indicative of the intensification of political dialogue that has occurred since late 2008, meaning in many cases the first visit by a Russian head of state and vice versa... [high-level political] visits have provided a suitable framework for the signing of agreements with the intention of promoting cooperation in all areas of activity—particularly in the fields of trade, finance, and science and technology – with specific projects to be developed in the medium and long term.[82]

The CAS embodied in globalization disseminates a vernacular pluralism that consists of commonplace, foundational, implicit/explicit ordering assumptions and principles about international organization.[83] The vernacular pluralism of globalization and its accouterments embody an overriding and stalwart belief in the existence of a universal, shared international or global interest among all members of the global community. Within the context of a CAS, CCNs function as a form of cultural homogenization on a structural level. As Daniel Egan notes, the Gramscian notion of hegemony, the levels of "civil society" and "political society," distinct states of affairs when discussing the state, become fused; civil society "is the ensemble of organisms commonly called 'private,' and that of 'political society' or the 'State.' "[84]

DOI: 10.1057/9781137308139

CCNs facilitate global governance via the functions they assume in a networked CAS. In a society of states, CCNs are *purveyors of information*, and wield influence and power on the world stage via a capacity to efficiently and effectively research, collect, analyze, and apply information more so than the cumbersome state bureaucracies. CCNs are able to generate in-demand information for state clients and private global market actors, for example, in the realms of finance, legal issues, and security alliances. CCNs become global information search engines that facilitate international relations in a manner consistent with a hyper-connected world. Information thus becomes a form of soft power. In addition to information generation, CCNs provide *forums or venues* that provide spaces and places for states to interact cooperatively, and are also *institutionalized normative agents* that express globalist values of world order and help establish *rules of engagement and rule implementation* among nation-states, for example, conventions and treaties. Lastly, within the context of a CAS, CCNs provide the fodder for global administration. International policies, agreements, trade relations, laws, and cooperative security arrangements—all of these states of affairs require administrative apparatuses that are specialized in various international state-tasks. *Specialization* is thus another complex element of CCNs that further facilitate international relations, as Russia–Latin America relations will demonstrate.[85]

Notes

1   Barry Buzan, "From International System to International Society of States: Structural Realism and Regime Theory Meet the English School," *International Organization*, Vol. 47, No. 3 (1993), pp. 327–352, and 331.

2   Hedley Bull, *The Anarchical Society: A Study of Order in World Politics*. Forward by Andrew Hurrell and Stanley Hoffman, 3rd ed. (Colombia University Press: New York, 2002), pp. 9–10.

3   Helga Turku, *Isolationist States in an Interdependent World* (Burlington: Ashgate, 2009), p. 37.

4   Ari Afilalo and Dennis Patterson, "Statecraft, Trade and the Order of States," *Chicago Journal of International Law*, Vol. 6, No. 2 (Winter 2006), pp. 725–759, especially, see p. 730.

5   Ibid., p. 731.

6   Turku, *Isolationist States in an Interdependent World* (Burlington: Ashgate, 2009), p. 1. Also see Sheila L. Croucher, *Globalization and Belonging: The Politics of Identity in a Changing World* (Lanham: Rowman & Littlefield, 2004).

DOI: 10.1057/9781137308139

7   Turku, *Isolationist States in an Interdependent World*, p. 2. For in-depth and comprehensive discussions on globalization, see Jacques Baudot, *Building a World Community: Globalization and the Common Good*, Royal Danish Ministry of Foreign Affairs Copenhagen in association with the University of Washington Press Seattle and London, (Seattle: University of Washington Press, 2001); Ino Rossi, (ed.), *Frontiers of Globalization Research: Theoretical and Methodological Approaches* (New York: Springer, 2007); David A. Deese, *World Trade Politics: Power, Principles, and Leadership* (New York: Routledge, 2008); Marshall R. Singer, *Weak States in a World of Powers: The Dynamics of International Relationships,* (New York: Macmillan, 1972); John Gerard Ruggie, *Constructing the World Polity: Essays on International Institutionalization* (New York: Routledge, 1998); Samuel Huntington, *The Third Wave: Democratization in the Late Twentieth Century* (Norman: University of Oklahoma Press, 1991), pp. 46–72; Ernst Haas, *Where Knowledge is Power: Three Models of Change in International Organizations* (Berkeley: University of California Press, 1990); Joseph Nye, "Neorealism and Neoliberalism," *World Politics*, Vol. 40 (1988), pp. 235–251.

8   Noam Chomsky, "Noam Chomsky Chats with Washington Post Readers," *The Washington Post*, March 24, 2006, accessed on 11 April 2008 <http://www.chomsky.info/debates/20060324.htm>.

9   Mark A. Smith, "Russia and Latin America: Competition in Washington's 'Near Abroad'?" Paper prepared for Research and Assessment Branch, *Defense Academy of the United Kingdom*, August 9, 2009 (United Kingdom), pp. 1–22, particularly, p. 1.

10 Anthony Giddens, *The Consequences of Modernity* (Cambridge, Oxford: Polity Press, 1990), p. 64.

11 Karl Marx and Frederick Engels, *The Communist Manifesto, A Modern Edition* (New York: Verso, 1998), p. 39.

12 Turku, *Isolationist States in an Interdependent World*, p. 3. Also see Jane Jenson and Boaventura de Sousa Santos, "Introduction: Case Studies and Common Trends in Globalization," in Jane Jenson and Boaventura de Sousa Santos, (eds), *Globalizing Institutions; Case Studies in Regulation and Innovation,* (Aldershot: Ashgate, 2000); Filippo Cesarano, *Monetary Theory and Bretton Woods: The Construction of an International Monetary Order*, Series: Historical Perspectives on Modern Economics (New York: Cambridge University Press, 2006); Robert C. Dash, "Globalization for Whom and for What," *Latin American Perspectives*, Vol. 25, No. 6 (1998), pp. 52–54.

13 See Philip Bobbitt, *The Shield of Achilles* (New York: Knopf, 2002), pp. xxii, 677–714.

14 Turku, *Isolationist States in an Interdependent World*, p. 4. Also see Peer C. Fiss and Paul M. Hirsch, "The Discourse of Globalization: Framing and Sense- Making of an Emerging Concept," *American Sociological Review*, Vol.

DOI: 10.1057/9781137308139

70 (2005), p. 42; Anne-Marie Slaughter, *A New World Order* (Princeton: Princeton University Press, 2004); David Held, Tony McGrew, Jonathan Perraton, and David Goldblatt, *Global Transformations* (Cambridge: Polity, 1999), p. 7; Sol Piccotto, "Networks in International Economic Integration," *Northwestern Journal of Law and Business*, Vol. 17 (1996–1997), p. 1014; Wolfgang H. Reinicke, "Global Public Policy," Foreign Affairs Vol. 76 (1997), p. 137; Martin Hollis and Steve Smith, *Explaining and Understanding International Relations* (New York: Oxford University Press, 1990), p. 83.

15 Walter Walle, "Russia Turns to the South for Military and Economic Alliances," *Council on Hemispheric Affairs*, May 8, 2012, 22 July 2012 <http://www.coha.org/russia-turns-to-the-south-for-military-and-economic-alliances/>.

16 Turku, *Isolationist States in an Interdependent World*, p. 5; and Katherine M. Franke, "The Domesticated Liberty of Lawrence v. Texas," *Columbia Law Review* Vol. 104, No. 5 (2004), pp. 1399–1426.

17 Walter Walle, "Russia Turns to the South for Military and Economic Alliances."

18 See Ray Kiely, *The Clash of Globalizations: Neo-Liberalism, the Third Way and Anti-Globalization* (Haymarket: New York, 2009).

19 Walter Walle, "Russia Turns to the South for Military and Economic Alliances."

20 Pravda.ru, "Russia Offers Latin America to Combat Drugs Together," March 12, 2012, accessed on 22 July 2012 <http://english.pravda.ru/russia/economics/12–03–2012/120748-russia_latin_america_drugs-0/#>.

21 Margaret E. Keck and Kathryn Sikkink, *Activists Beyond Borders: Advocacy Networks in International Politics* (Ithaca: Cornell University Press, 1998), p. 2.

22 Alfred North Whitehead, *Symbolism: It's Meaning & Effect* (New York: Fordham University Press, 1927), p. 63.

23 Sistema Económico Latinoamericano y del Caribe, "Final Report of the Regional Meeting on Recent Developments in Economic Relations between the Russian Federation and Latin America and the Caribbean," Caracas, May 21, 2012, accessed on 22 July 2012 <http://www.sela.org/attach/258/default/Final_Report_Recent_Developments_in_Economic_relations_between_the_Russian_LAC.pdf>.

24 Ibid.

25 Margaret P Karns and Karen A. Mingst, *International Organizations: the Politics and Processes of Global Governance* (Boulder, Colorado.: Lynne Rienner Publishers, 2010), p. 59.

26 John Bolton, "Should We Take Global Governance Seriously?" *Chicago Journal of International Law*, Vol. 1 (2000), p. 206. For examples of NGOs seeking to substantively influence and alter a state-centric paradigm of international organization, see the recent case of the Free Gaza Movement's

DOI: 10.1057/9781137308139

clash with the state of Israel in Isabel Kershner, "Deadly Israeli Raid Draws Condemnation," *New York Times* May 31, 2010, accessed on 9 June 2010 <http://www.nytimes.com/2010/06/01/world/middleeast/ 01flotilla.html>.

27   Martin Shapiro, "Administrative Law Unbounded: Reflections on Government and Governance," *Indiana Journal of Global Legal Studies* 8 (2001), p. 369.

28   Bloom, *Global Brain: The Evolution of Mass Mind from the Big Bang to the 21st Century* (John Wiley: New York, 2001)., p. 27.

29   Karns and Mingst, *International Organizations: the Politics and Processes of Global Governance*, p. 217.

30   Jean Francois Lyotard and Jean-Loup Thebaud, *Just Gaming*, trans. Wlad Godzich (Minneapolis: University of Minnesota Press, 1999), p. 28.

31   Bloom, *Global*, p. 30.

32   Ibid.

33   Ibid.

34   Ludwig Wittgenstein, *Tractatus Logico-Philosophicus*, trans. D. F. Pears and B.F. McGuiness. (London: Routledge, 1975), p. 68.

35   Ibid., p. 82.

36   Sistema Económico Latinoamericano y del Caribe, "Recent developments in Economic Relations between the Russian Federation and Latin America and the Caribbean: Institutional and Cooperation Mechanisms for Strengthening Relations."

37   This discussion is based on Bloom, *Global*, p. 35.

38   Roy J. Eidelson, "Complex Adaptive Systems in the Behavioral and Social Sciences," *Review of General Psychology*, Vol. 1, No. 1 (1997), pp. 42–71. See particularly, p. 43.

39   See Richard K. Betts, *Conflict After the Cold War*, Updated Edition (2nd ed.) (Longman: New York, 2004).

40   Bloom, *The Genius of the Beast: A Radical Revision of Capitalism* (Amherst: Prometheus, 2010), p. 373.

41   See Michel Foucault, *Archaeology of Knowledge and Discourse of Language*, trans. A.M. Sheridan Smith (New York: Pantheon, 1972). See, p. 38.

42   The discussion that follows on the components of the learning machine is drawn from Howard Bloom, "Beyond the Super-computer: Social Groups as Self-Invention Machines," *Sociobiology and Bio-politics*, Albert Somit and Steven A. Peterson (eds), *Research in Bio-politics*, Vol. 6 (Greenwich: JAI Press, 1998), pp. 43–64.

43   See David Luban, "A Theory of Crimes Against Humanity," *Yale Journal of International Law*, Vol. 29 (2004), pp. 85–167.

44   Roy J. Eidelson, "Complex Adaptive Systems in the Behavioral and Social Sciences," *Review of General Psychology*, Vol. 1, No. 1, (1997), pp. 42–71. See pp. 43–44.

DOI: 10.1057/9781137308139

45 H.A. Simon, "Near Decomposability and Complexity: How a Mind Resides in a Brain," in H. Morowitz and J. L. Singer (eds), *The Mind, the Brain, and Complex Adaptive Systems* (Reading, Massachusetts: Addison Wesley; Santa Fe Institute Studies In the Sciences of Complexity, Proceedings, 1995), Volume XXII, pp. 25–43. See especially p. 26.

46 Eidelson, "Complex Adaptive Systems in the Behavioral and Social Sciences," p. 44.

47 Bloom, *Global*, p. 49.

48 For a more detailed account, see Howard Bloom's discussion of memes in *Global Brain,* Chapter 2.

49 Eidelson, "Complex Adaptive Systems in the Behavioral and Social Sciences," p. 44.

50 Richard Dawkins, *The Selfish Gene* (New York: Oxford University Press, 1976), p. 206.

51 See Friedrich Nietzsche, "On Truth & Lying in a Non-Moral Sense," trans. Ronald Spiers, in Vincent B. Leitch, William E. Cain, Laurie A. Finke, Barbara E. Johnson, John McGowan and Jeffrey J. Williams (eds), *The Norton Anthology of Theory & Criticism* (New York: WW Norton, 2001), p. 875.

52 Ibid.

53 Francisco J. Varela, "Autonomy and Autopoiesis," in Gerhard Roth and Helmut Schwegler (eds), *Self-Organizing Systems: An Interdisciplinary Approach*, (New York: Campus Verlag, 1981), p. 15.

54 Humberto R. Maturana, "The Organization of the Living: A Theory of the Living Organization," *International Journal of Man-Machine Studies* Vol. 7, No. 3, (May 1975), pp. 313–332. See especially, p. 316.

55 Bloom, *Global*, p. 71.

56 Ibid., p. 84.

57 Oscar Montealegre, "Brazil's Trade Diplomacy," *Diplomatic Courier*, January 20, 2011, accessed on 30 July 2012 <http://www.diplomaticcourier.com/news/bric/37>.

58 Ibid.

59 Bloom, *Global*, p. 84.

60 See Eric J. Hobsbawm, *Nations and Nationalism Since 1780* (New York: Cambridge University Press, 1990).

61 Bloom, *Genius*, p. 295.

62 Sistema, "Recent.".

63 Gilles Deleuze and Felix Guattari, "Kafka: Toward a Minor Literature," trans. Dana Poland, in Vincent B. Leitch, William E. Cain, Laurie A. Finke, Barbara E. Johnson, John McGowan and Jeffrey J. Williams, (eds), *The Norton Anthology of Theory & Criticism* (New York: WW Norton, 2001), p. 1599.

64 Bloom, *Global*, p. 111.

65 Bloom, *Genius*, p. 303.

DOI: 10.1057/9781137308139

66   Jean Francois Lyotard, *The Postmodern Condition: A Report on Knowledge*, forward by Fredric Jameson, trans. Geoff Bennington and Brian Massumi (Minneapolis: University of Minnesota Press, 1984), p. 5.

67   Stanford Encyclopedia of Philosophy, *Jean Bodin*, March 25, 2005, revised 14 June 2010, accessed on 12 January 2011 <http://plato.stanford.edu/entries/bodin/#4>.

68   Generation Online, *Jean Bodin: On Sovereignty*, no date, accessed on 12 January 2011 <http://www.generation-online .org/p/fpbodin1.htm>.

69   The discussion that follows about strategic thought and interests is from Marvin L. Astrada, *Strategic Culture: Concept and Application*, Applied Research Center, Florida International University, Miami, Florida, February 2010, pp. 5–6, accessed on 18 January 2011 <http://strategicculture.fiu.edu/Approach/StrategicCultureConceptandApplication.aspx>.

70   *US DOD Dictionary of Military Terms* <http://www.dtic.mil/doctrine/jel/doddict/> March 17, 2009, accessed on 15 July 2009.

71   Ibid.

72   Ibid.

73   Martin Hollis and Steve Smith, *Explaining and Understanding International Relations* (Claredon: Oxford, 1990), p. 83.

74   For discussions of how each of the aforementioned concepts impact sovereignty in a system of states paradigm, see Jean Bodin, *On Sovereignty: Four Chapters From the Six Books of the Commonwealth,* ed. and trans. Julian H. Franklin (Cambridge: Cambridge University Press 1992); J. A. Hall, *States in History* (New York: Blackwell, 1986); Joseph R. Stayer, *On the Medieval Origins of the Modern State* (Princeton: Princeton University Press, 1970); Nicholas Onuf, "Sovereignty: Outline of a Conceptual History," *Alternatives* Vol. 16, No.4 (1991), pp. 425–446; Janice Thomson, "State Sovereignty in International Relations: Bridging the Gap between Theory and Empirical Research," *International Studies Quarterly* Vol. 39, No.2 (1995), pp. 213–233; F. H. Hinsley, *Sovereignty*, 2nd ed. (Cambridge: Cambridge University Press, 1986); and Andreas Osiander, "Sovereignty, International Relations, and the Westphalian Myth," *International Organization* Vol. 55, No. 2 (2001), pp. 251–287.

75   Bloom, *Global*, p. 118.

76   Bennett Stark, "A Case Study of Complex Adaptive Systems Theory Sustainable Global Governance: The Singular Challenge of the Twenty-first Century," University of Ljubljana, WISDOM RISC-Research Paper No. 5, July 2009, pp. 1–38. See p. 19.

77   Bloom, *Global*, p. 118.

78   Bloom, *Global*, pp., 119–120. See Balkrishna Govind Gokhale, *Asoka Maurya* (Twaynes: New York, 1966), p. 56, and pp. 79–80.

79   Bloom, *Global*, p. 120.

DOI: 10.1057/9781137308139

80   Bloom, *Global*, p. 129.

81   Ibid., p. 186.

82   Sistema, "Recent."

83   See Daniel Egan and Levon A. Chorbajian, (eds), *Power: A Critical Reader* (Upper Saddle River: Pearson, 2005), pp. xvii–xviii; Institute for the Future, "The Cooperation Project: Objectives, Accomplishments, And Proposals," March 30, 2005, accessed on 20 May 2010 <www.rheingold.com/ cooperation/CooperationProject_3_30_05.pdf>.

84   Daniel Egan and Levon A. Chorbajian, *Power: A Critical Reader*, p. 10.

85   See Rodney Bruce Hall, "Private Authority: Non-state Actors and Global Governance," *Harvard International Review,* June 22, 2005, accessed on 19 April 2010 <http://www.allbusiness.com/public-administration/national-security-international/462542–1.html>.

DOI: 10.1057/9781137308139

3

Exploring the Emergent State-Society Synthesis: Russia–Latin America Relations

Abstract: *Russia–Latin America relations are an empirical manifestation of how state power, perceptions, interests, and methods for realizing those interests have been impacted by the hyper-networked nature of contemporary international order and organization. Russia–Latin America relations illustrate how network-based cooperative frameworks and institutions both complement and challenge state power and prerogatives on the world stage. Additionally, Russian engagement with the region illustrates the consequences that devolve from a complex, hyper-networked international order. It appears that networks facilitate engagement and are having an independent effect on the potential for constructive and integrative engagement that lies outside the traditional sphere of action and limitations under a system of states.*

Astrada, Marvin L. and Martín, Félix E. *Russia and Latin America: From Nation-State to Society of States*. New York: Palgrave Macmillan, 2013. DOI: 10.1057/9781137308139.

In the present international order, the system of states has undergone a profound reconfiguration due, in large part, to the rise of CCNs. Within an evolving global milieu, Russia–Latin America relations exemplify this profound reconfiguration. Specifically, Russia–Latin America relations are an empirical manifestation of how state power, perceptions, interests, and methods for realizing those interests have been impacted by the hyper-networked nature of contemporary international organization. The use and importance of various CCNs by states for the realization of various motives, e.g., military and economic, will be examined in the context of Russia–Latin America relations to illustrate how network-based cooperative frameworks and institutions both complement and challenge state power and prerogatives on the world stage, and the consequences that devolve from a complex, hyper-networked international order. Russia's engagement with the region illustrates "the expansion of [complex economic, political, and] social networks that facilitate the reproduction of transnational... economic organization, and politics."[1]

This chapter will provide a general analysis of select themes and issues that inform Russia–Latin America relations and the subsequent chapter will examine these themes in greater detail. At the outset, it is important to note that *realpolitik* considerations undergird Russia's non-military (and military) involvement in Latin America. Soft power CCNs, however, which are being employed to facilitate engagement are having an independent effect on the potential for constructive and integrative engagement that lies outside the traditional sphere of action and limitations under a system of states. Russian policy, as with other middle (and great) powers, is part of a strategic global agenda that transcends the realization of economic self-interest. In the process of trying to realize its strategic goals, however, Russia's utilization of soft power and cooperative networks to implement policy is contributing to the emergence of society and alternative forums and tools for interstate relations. For the region, Russian engagement presents opportunities to network outside what has been historically referred to as America's sphere of influence. This is the case because transnational engagement generates complex spaces of interaction between international actors.

> Constructing transnational political spaces should be treated as the resultant of separate, sometimes parallel, sometimes competing projects at all levels of the global system—from the 'global governance' agenda of international organizations and multinational corporations to the most local 'survival strategies,' by which transnational networks are socially constructed.[2]

DOI: 10.1057/9781137308139

Although Russia's approach to the region may not be as fully developed as its policy and approach to Europe, Russia and Latin America are pursuing agendas via intensive and extensive CCNs rooted in formal (legal) agreements based on trade and investment in the broadest sense of the terms. According to the Russian government, developing and enhancing CCNs with the region is a cornerstone of its long-term foreign policy interest. Russian engagement is based on pragmatic, strategic, economic, and non-military or security concerns (e.g., strengthening civilian travel and cultural exchange ties) that reflects the "closeness" of Russian-Latin American "views on key issues in world politics and economics," with CCNs being key to facilitating "collective and concerted action with regard to the generally recognized norms of international law, cultural and historical traditions."[3]

Of consequence for the region, in addition to the establishment of CCNs in energy, resource exploration, space technology, trade, tourism, military-to-military partnerships, military ordinance, and Russia's strategic vision—unlike the previous Soviet strategic vision—is based on fostering a "multi-polar world" (in contrast to the bipolar structure that dominated global affairs from 1945–1992). Russia's cooperative partnerships with Latin America constitute a "concrete step in laying the cornerstone of a new geopolitical order based on multilateral values, a New World Order."[4] Multipolarity, in large part, is based on societal notions of world order and the use of CCNs to facilitate a multipolar world, since multipolarity bases international relations on non-coercive engagement. Given the aim of Russia's foreign policy, its approach to Latin America thus has significant consequences for the region.

In general, Russian strategy in Latin America has been viewed by some commentators as a

> geopolitical approach directed against the U.S. with an economic component, rather than an economic approach to foreign policy with strategic objectives... Moscow poses [a threat] to the region... from its weapons sales to Venezuela, which the latter is using in support of insurgency in Colombia if not elsewhere.[5]

Yet, Russian engagement can also be viewed from a soft-power perspective that entails additional non-military-security objectives. In addition to military-security threats, Russian engagement on the soft-power level may be viewed as part of a strategy to present the region with an alternative major trading partner. Russia competes with

DOI: 10.1057/9781137308139

China and the US, among other states for Latin American markets. This undermines tacitly the "legitimacy" of having there being a single dominant power outside of Latin America and the Caribbean. That is, he US. Russian engagement via CCNs may be viewed as effectively utilizing soft-power resources and networks to complement any military-security in the region via diplomacy, trade, investment, security alliances, energy, and other types of cooperative partnerships. Since 2000 Russia gradually restored its presence in the region, integrating itself into the region's affairs related to commercial, legal, technological, political, cultural, and in resource development and exploitation activities.[6]

Russian involvement in Latin America can be viewed as a manifestation of Russia's long-term security agenda—one that uses soft power, CCNs, and societal notions of international order. While there are indeed *realpolitik* considerations in Russian engagement, it seems short-sighted to limit Russian motives (as well as the consequences of implementing CCNs to facilitate engagement) to the limited logic of the states system. Multipolarity and extensive, complex, cooperative agreements between the region and Russia also reflect a genuine shift in perception and behavior that reflects the changed global landscape—a landscape that is now characterized by a multiplicity of networks facilitated by non-state actors defined by highly complex interconnectivity and interaction. Russian engagement can thus be viewed as a societal phenomenon, in that the proliferation of CCNs to conduct interstate relations, even in light of there being military-security interests at play, suggests a partial departure from a states system logic and the opening up of alternatives for state engagement based on CCNs. Relegating oneself to the traditional logic of state perception and interests has the effect of overlooking important soft-power explanatory factors that have significant consequences for conceptualizing international relations and organization.

The one soft power resource that the Soviet Union employed, i.e., ideology, while not as integral as it was to Soviet foreign policy, has presented Russia with some inroads into the region. An ideology rooted in free trade and universal values, however, and not in superpower politics, seems to be at play in present Russia–Latin America relations. Russia has successfully managed to reinsert itself substantively into Latin America over the last 15 years. As Nil Nikandrov observes, Russia's recent success in Latin America has been due to, in part, the creation of

DOI: 10.1057/9781137308139

favorable conditions for the development of relations with the countries of the continent. The ideological component does not matter here—Moscow is ready to cooperate with the countries which have left-socialistic and right conservative governments....In 2005–2009 Russia's trade turnover with the countries of Latin America increased to $18 billion from about $6 billion. Some experts name the achievements of the Russian policy in Latin America "the most efficient outcome of Moscow's international activities in recent years," "momentous," and in many ways contributing to "the modernization of the regional relations."[7]

Russia retains relations with its former (when it was the USSR) as well as "new" ideological partners such as Venezuela's Hugo Chávez, Nicaragua's Daniel Ortega, Ecuador's Rafael Correa, then-president Luiz Inácio Lula da Silva in Brazil, Néstor Kirchner and Cristina Fernández in Argentina, Tabaré Vázquez in Uruguay, Bolivia's Evo Morales, and the Castro brothers in Cuba.[8] Moscow has capitalized on this phenomenon to enhance its relations with various Latin American states, utilizing political ideology—even if only rhetorically or symbolically—and common interests and credit lines (e.g., Bolivia receiving a $100 million credit line to upgrade its armed forces) in order to improve and establish ties.[9] Even states that are not anti-American *per se*, such as Peru, continue to have substantive ties with Russia, as the Andean state's military equipment is of Soviet origin (helicopters, Antonov and Sukhoi planes) and it continues to rely on Russia as a weapon supplier.

Russia's engagement with the region coincides with Latin American countries' efforts to diversify their interests via CCNs. Russia is one of several countries competing for the region's attention. In the context of multilateralism and the networked nature of international organization, expansion of contacts via CCNs makes cooperation more substantive, more expansive, and more complex. Granted that hard power considerations remain relevant: for instance, Nicaragua recognizing Abkhazia and South Ossetia to obtain the promise of Russian military and economic assistance to replace its aging arsenal of Soviet-era weaponry; Cuba's continuing interest in strengthening ties based on a military as well as non-military interests; Venezuela's use of Russian military support to maintain the regime in power; and Guatemala's Vice President Rafal Espada expressing interest in acquiring Russian arms in exchange for food—"Guatemala is interested in acquiring planes, armored vehicles and other arms to struggle against organized crime in the country. We could pay for the arms with coffee and sugar," Espada said.[10] Yet, soft

DOI: 10.1057/9781137308139

power and the employment of CCNs by Russia assume an important role and do have an independent effect on the dynamics and conduct of interstate relations.

The Cold War, the USSR, and Russia

As mentioned above, Russia's involvement in the region is not a "new" policy. When assessing Russia–Latin America relations and the impact of Russia on the region in the present, however, there has been a profound shift in the "why" and "how" of engagement with the region. Some degree of historical continuity can be observed between the macro-level motives that informed the USSR's foreign policy and Russia's current engagement with the region, for example, the role of ideology in framing engagement strategies. The methodology, however, has changed, and the motives for engagement have been significantly modified. While the USSR's engagement with the region was premised on a purely system of states perspective, Russia's present-day engagement has been complicated, modified, and challenged by the onset of full-blown globalization and the rise of CCNs in the organization of international affairs. Before proceeding to an analysis of the changes that have occurred in the conceptual, perceptual, and organization of interstate relations between Russia and the region, the following section will provide a cursory account of the modern historical basis of Russia–Latin American relations and Russian foreign policy toward the region. A comparative analysis better allows for fleshing out the changed nature and character of present Russia–Latin America relations.

Historical continuity between the modern and post-modern Russian state involves the projection of power and the augmentation and enhancement of state power. Key differences between the past and the present is that Russia is no longer a superpower, and that a virulent political ideology (communism vs. capitalism) dividing the world into opposing camps as an ordering principle of interstate relations has given way to a more cooperative, integrated, and networked international order that relies upon soft power and CCNs as the primary form of interstate relations. With the loss of political ideology as a mainstay of Russian engagement with the region, numerous opportunities have arisen for Russia and the region's countries to benefit, jointly and severally, in the realms of economic development, trade, and political stabilization.

DOI: 10.1057/9781137308139

The Cold War (1945–1989) explicitly informed the USSR's (1917–1991) perceptions, policies, and engagement with Latin America. Soviet strategic and political interests were entrenched in a binary ideological contest (inextricably enmeshed with power politics) that pitted communism vs. capitalism. Superpower politics, in particular the presence of a state-centric bipolar balance of power system premised on the threat of annihilation via weapons of mass destruction (WMD), underlay the ideological, geopolitical, and economic dimensions of international order, and structured and impacted the internal socioeconomic and political development and wellbeing of Latin America. Political ideology was an operative consideration in the conduct of international relations. Cold War politics were steeped in ideology and *realpolitik* considerations and were a mainstay of interstate engagement.[11] Geopolitically, Latin America was viewed as providing the USSR with the capacity to establish a presence in US's sphere of influence. As it has been characterized in several sources,

> Latin America has historically played an important role in the struggle for sphere of influence between the U.S. and Russia. Throughout the Cold War, the Soviet Union...paid special attention to intensifying relations with countries in the U.S.'s backyard as a response, [among other considerations], to U.S. interference in regions commonly known to be under Soviet influence.[12]

In sum, engagement was premised on military considerations and effectuating a change in the global balance of power in favor of the Soviet state.

Soviet political ideology and foreign policy objectives

Two fundamental goals of Soviet foreign policy remained constant throughout the Cold War: first, strategic considerations rooted in superpower politics enmeshed in an ideological and militarized contest, and, second, regional influence based on strategic interests. During the latter half of the twentieth century, Soviet foreign policy placed primary emphasis on leveraging US global and regional power—the US being perceived as the most serious threat to the USSR's national security and aspirations for regional hegemony and global superpower status. Additionally, as reported in the US Library of

DOI: 10.1057/9781137308139

Congress's "Country Study: Soviet Union," the Soviet Union, among its remaining foreign policy goals, highly prioritized its relations with Eastern Europe and Western Europe, and it gave very little priority to Latin America and the Caribbean, except, insofar as the region "either provided opportunities for strategic basing or bordered on strategic naval straits or sea lanes."[13]

In the initial phases of the USSR's development during the first-half of the twentieth century, Latin America, generally speaking, had ranked quite low on the list of Soviet foreign policy priorities. Yet, from the late-1950s on, a Soviet presence in Latin America had grown steadily but slowly, in large part due to anti-US ideological movements and anti-US sentiments from Latin America. Until the Khrushchev period and the forging of relations with communist Cuba, Latin America was generally regarded as being well within US sphere of influence, or as its "backyard." Up until the early 1960s, the USSR had "little interest in importing Latin American raw materials or commodities, and most Latin American governments, traditionally anti-communist, had long resisted the establishment of diplomatic relations with the Soviet Union."[14] A major event that initiated a turning point in USSR–Latin America relations was the Cuban Revolution of 1959, in which opposition forces led by Fidel Castro toppled the US-backed government of Fulgencio Batista. Fidel Castro proceeded to gradually define revolutionary Cuba as a communist state, and developed close ties with the USSR to the point that Cuba became the recipient of massive Soviet military and financial aid in exchange for a Soviet strategic presence on the island in the form of a listening post at Lourdes and as a resupply base for Soviet strategic and long-range bombers and naval vessels engaged in some sort of off-shore balancing. Further, by 1965, revolutionary Cuba was well entrenched in the Soviet camp, and provided the USSR with something unprecedented, i.e., an ideological and military foothold in the Americas.

Cuba initially advocated armed revolutionary struggle as the only viable and legitimate method by which to effectuate political change in Latin America. "However, after armed struggle failed to topple the governments of the Dominican Republic and Venezuela in the early 1960s, the Soviet leadership stressed the 'peaceful road to socialism'[;] this path involved cooperation between communist and leftist movements in working "for peaceful change and electoral victories."[15] It has been argued elsewhere, however, that the armed struggle and guerrilla-

DOI: 10.1057/9781137308139

warfare strategy was truly a Cuban approach to counterbalance Soviet relations with the US.[16] The Cubans feared, the argument goes, that the Soviets might have reached a secret pact with the Americans, particularly after the Missile Crisis and the failed US-led Bay of Pigs invasion, leaving them exposed and an easy target of another potential US-led invasion. Thus, the armed struggle or guerrilla foci strategy was used by the Cubans as a sort of international diplomatic blackmail tool against the Soviets. This was particularly clear, later in 1968, when the Soviets invaded Czechoslovakia, and the Cubans dragged their feet for several weeks before lending formal political and ideological support to the Soviet action in Eastern Europe. The Cubans "hoped" that the Soviets would come to their rescue in case the US tried to topple the Castro regime again. In retrospect the historical record shows that this was an unfounded fear on the part of the Cuban leadership since the US and the Soviets had negotiated a secret non-aggression "understanding" as part of the 1962 Kennedy–Khrushchev Pact, ending the Cuban Missile Crisis. In addition to other agreements like establishing a hotline between the Kremlin and Washington and allowing the Soviets to keep a combat brigade on Cuban soil, the Americans committed to a non-aggression policy against the Castro regime.

After 1973, however, with the election and subsequent overthrow of Salvador Allende's socialist government in Chile, the USSR proceeded to provide a massive military support to Cuba and to other leftist-insurgent and left-leaning groups, i.e., anti-American governments in Latin America. This strategy was also highly influenced by the negative outcome of the 1973 Yon Kippur War against Soviet allies in the Middle East. Evidently, after the failures in the Middle East, the Soviets wanted to "reassure" its allies worldwide of their prowess and commitment. Latin America and the Caribbean were no exception, and the Soviets proceeded to build closer and more committed relations with their allies in the western hemisphere. Accordingly, superpower politics and interests influenced the USSR's decision to use Cuba as a touchstone for disseminating arms, aid, and ideology based on anti-Americanism throughout Central America.[17]

Massive aid, historically and in the present, is not the same as full-fledged financing of a country's economy. Soviet investment—as is the case with the contemporary Russian State—was deemed rational, strategic, and designed to procure apperceived benefits vis-à-vis Soviet interests. Investment, however, was premised on ideology—a most

DOI: 10.1057/9781137308139

irrational basis upon which to premise policy and strategic interest. Nevertheless, investment, past and present, has been premised on utilizing hard- and soft-power resources for empowering the state. The effects of former Soviet and, now, Russian investment and engagement has produced miscellaneous results in the region, from entrenchment of virulently anti-American regimes (e.g., Castro's Cuba) to the support of revolutionary movements (Sandinista Nicaragua), to fostering a variety of CCNs premised on mutual material and security gains via cooperative and sophisticated engagement.

In South America, the USSR engaged countries via "soft power" resources, viz., diplomatic relations and trade/investment agreements. Peru was the only state to engage in substantial arms purchases and military-to-military relations with the USSR. In February 1969, Peru established diplomatic relations with the USSR; in 1973, Soviet arms arrived in Peru with major transfers occurring after 1976, e.g., fighter-bombers, helicopters, jet fighters, surface-to-air missiles, and other sophisticated weaponry being imported.[18] In contemporary world affairs, Russia–Peru relations remain stable due to material and strategic interests that transcend political ideology. Russia has utilized this approach to engage non-ideological regimes in the region. This is a dynamic that must be kept in mind when assessing Russia–Latin America relations, i.e., the fact that ideology and explicitly engaging the US in a hostile manner by using countries in the Americas as proxies are no longer key reasons for Russian rapprochements with the region. Present Russia–Peru relations are characterized by a

> positive evaluation of dynamically developing relations in the sphere of military and technical cooperation and preventing disaster situations... [Each country has] reaffirmed the commitment of both countries to universally recognized principles of international law, strengthening of multilateral diplomacy, central coordinating role of the UN in world affairs and countering new threats and challenges, including combating terrorism and drug traffic.[19]

Soviet policy—as is presently the case with Russian policy, although without the explicitly primary focus on use of soft power CCNs—toward Latin America was not monolithic; different tactics and tracks were utilized to more effectively leverage American power in the region. The Soviet *modus operandi* was to support military actions, subversive movements, and revolutionary violence in Central America while pursuing

DOI: 10.1057/9781137308139

limited engagement with South America (with the notable exception of Peru). As reported in the US Library of Congress Country Study of the Soviet Union, the

> range of instruments of influence used in Central America and South America, while varying in their mix over time, nevertheless indicated that all instruments, including support for subversive groups and arms shipments to amenable governments, had been used in Central America and South America in response to available opportunities, indicating shifting emphases but a basically undifferentiated policy toward Latin America.[20]

Russian foreign policy and Latin America

The relatively unexpected dissolution of the USSR in 1991 ushered in profound consequences for international order and the system of states. For nearly five decades preceding the Soviet collapse, international organization had been structured along a politicized *realpolitik* bipolar axis; the abrupt change in the fabric of international affairs affected all regions of the world. As Russia turned inward to deal with its internal post-communist transition, former USSR allies in Latin America were left to fend for themselves—most notably Cuba. "For Latin America, the 1990s have been characterized as the 'lost decade,' due to the turmoil and tensions that consumed the region's attention as a result of Russia's turn inward."[21] Shortly after the demise of the USSR, the entire Eastern Bloc, the constituent components of the fabled "Iron Curtain," de-communized, leaving Cuba and the rest of Latin America without the massive military, trade, and/or financial support that had flowed into the region for nearly five decades based on Cold War superpower politics. Russia was thus forced, as during the early days of the Bolshevik Revolution, to drop out of world affairs and focus exclusively on its chaotic domestic situation. Russia was no longer a superpower, no longer a global power player, no longer occupied the cardinal role it had as balancing US power in a bipolar world.

In the present international order, Russia has enthusiastically embraced soft power in the form of variegated CCNs as a primary way in which to engage Latin America. Russia's present foreign policy is guided by establishing complex and extensive networks of cooperation with the

DOI: 10.1057/9781137308139

countries of Latin America. Russia has established relations based on CCNs

> with all 33 countries of the region. With many of these countries Russia carries out mutually beneficial projects in such sectors as energy, high technologies, [and] infrastructure[, and] plans to enhance mutually beneficial ties with the organizations within [the region, such as] Mercosur (Southern Common Market) which comprises Argentina, Brazil, Uruguay, Paraguay, Venezuela... and also with the number of allied countries.[22]

Russia's engagement with the region, via economic and diplomatic CCNs, is the result of a policy initially articulated by Russian Foreign Minister Yevgeny Primakov (1997), while visiting Argentina, Brazil, Colombia, and Costa Rica. During his visits to the respective countries, Primakov stated that Russia should have ties with all continents and regions within the world.[23] Following this notion, Russia proceeded to initiate extensive and intensive ties with Latin America in the form of CCNs based on trade, investment, development aid, energy partnerships, security alliances and cooperation, arms sales, diplomatic relations, and promoting the view of Russia as an alternative source of political and economic support. Societal notions of interstate relations and order are immanent in Russia's attempt to enhance its power, as is the case with other middle powers such as China because of the cooperative and integrative nature of engagement. Indeed, the very notion of power—conception, perception, and application—has been deeply impacted by the employment of CCNs to realize state goals. As noted by Stephen Blank, in 2006, Russian Foreign Minister Lavrov declared that Russian policy vis-à-vis the region was a high priority for Russian foreign policy interests. Accordingly, he explains, quoting Foreign Minister Lavrov, that

> "the countries of Latin America and the Caribbean Basin (LACB) occupy an increasingly noticeable place in the system of contemporary international relations. Our contacts with them...are an important component of the international efforts of Russia in tackling the problems common to the entire world community"...Russia uses areas of comparative economic advantage (energy, arm sales, space launches, sales of nuclear reactors) to leverage political support for Russian positions in the region.[24]

Under the auspices of President Vladimir Putin, Russia forcefully remerged in the twenty-first century onto the world stage—not in the capacity of superpower, but as a global player that demanded attention

DOI: 10.1057/9781137308139

and that had an alternative approach to world order. In a speech delivered at the 43rd *Munich Security Conference*, Putin reasserted Russia's desire to participate in a new or rather alternative ordering principle for international affairs. As he explained,

> What is a unipolar world? No matter how we beautify this term, it means one single center of power, one single center of force and one single master... The U.S. has overstepped its borders in all spheres—economic, political and humanitarian, and has imposed itself on other states... We see no kind of restraint—a hyper-inflated use of force.[25]

Russia has characterized US power as a phenomenon in need of being countered, and rather than offer Russian power alone as a solution, the strategic vision proffered has been a "multipolar" world, in which power is "shared" by states for a collective benefit as opposed to being monopolized by a single state; hence, the need to dispense with singular state-centric notions of international order. This sentiment, despite regional countries' ideological positions or previous history of relations with the US and the USSR, has reverberated throughout the region, receiving a positive reception. Networks—CCNs—have, thus, been utilized to further Russian interests, but in the process such networks have been creating viable and powerful alternatives to the myopic focus of a system of states paradigm.

To accomplish its aim, Russia has sought to use Latin America as springboard for diversifying its interests, and for realizing a multipolar world. As Alex Sanchez notes,

> Russia's reemergence is occurring at a time when much of Latin America is striving towards a more Washington-free environment by promoting indigenous economic integration and looking in directions other than at the U.S. for its trade and political partners... and the desire of others to ally themselves with extra-hemispheric powers, (like the recently formed IBSA, consisting of South Africa, Brazil and India), also has contributed to opening up the environment for Russia's growing influence.[26]

Beginning in 2000, elections in Latin America brought various leaders, e.g. in Bolivia, Honduras, Ecuador, El Salvador, and Nicaragua, with an openness to engage and ally themselves with extra-hemispheric powers, and this development has created opportunities for Russia to expand its influence via extensive and intensive CCNs.

Russia has utilized soft power resources (in tandem with its hard power resources) in a world order that has been undergoing

DOI: 10.1057/9781137308139

recalibration and reconfiguration since the end of the Cold War. In 2001, Russian President Putin wrote a telegram to "participants in a conference on Latin America that political dialogue and economic links with the region were important and would be mutually beneficial. He cited the establishment of links in science, education, and culture as particular areas of focus."[27] Thus, Russia's involvement in the region exemplifies the tension and concomitant complementary effect of soft power and CCNs in the conduct of international relations, and the uses of societal notions based on a globalist paradigm to more efficiently and effectively facilitate the day-to-day business of states. The tension between system and society captures the complexity and challenges of the "effort to differentiate the local from the national and the global political organization of transnational spaces points to the growing interdependence of geographical scales."[28] Soft power is, therefore, not to be taken lightly nor dismissed out of hand. In general, Russian engagement and investment in the region has produced sundry effects, which include extra-regional states creating alternative means of economic development and political support for select countries in the region, ranging from Brazil and Argentina to Cuba and Nicaragua, and creating more powerful, diverse, and integrating networks of a complex and cooperative nature. Upon completion of his Latin American tour, Lavrov stated that,

> "Brazil is interested in our joining in major projects of interregional importance, including a transcontinental gas pipeline, and modernization of railways in the continent."... Russo-Brazilian political relations are also on the rise as referred in Moscow's support of Brazil obtaining a permanent seat on the United Nation's Security Council.[29]

As in the case of Brazil, Russia has actively sought out multifarious partnerships at the highest level with all the countries in the region, e.g., in the case of the Caribbean, "the number of Russian tourists visiting the Dominican Republic [has] greatly risen, and negotiations [are] underway regarding the opening of a Russian consulate general on the island."[30] On the political front, it is interesting to note how Russia has not fully ignored ideology as a means of facilitating a presence in the region. A recent visit to Cuba by Prime Minister Mikhail Fradkov is a case in point.

> Reflecting upon Fradkov's visit, Russian Deputy Foreign Minister Sergey Kislyak explained: "A period of adapting to new realities was not easy, but

DOI: 10.1057/9781137308139

now we are moving towards a new level of cooperation and mutual interaction with our Cuban friends…Cuba has been and will remain our high-priority partner in Latin America." He went on to add that Russia would continue to advocate "the abolition of the U.S. economic embargo and other sanctions against Cuba."[31]

More recently (July 20, 2012), President Putin and Minster Medvedev met with Cuban President Raúl Castro in Moscow to discuss Russian-Cuban relations. "Putin told Castro that Cuba is a traditional partner, noting the 110 year anniversary of diplomatic relations between the two nations. Medvedev said the two countries must exploit the potential of their relations carefully and rationally on both economic and humanitarian levels."[32]

Russia's diversification within the region revolves around extensive economic investment, diplomatic relations, arms sales, tourism, trade, military cooperation, and energy development. Such diversification has implications for fostering alternative avenues of engagement for state-to-state interaction and international organization. Trade, investment, development, finance, and humanitarian assistance efforts are forming complex and cooperative networks ensconced within an overarching systemic, i.e., globalized, context based on societal notions of order and organization. According to the Russian government, there is sustained interest in Latin America

> among the Russian business community, including in the sphere of energy and hydrocarbon production. Major Russian companies…and a number of others work in Mexico, Venezuela, Bolivia, Colombia, Guyana and Cuba. Negotiations are under way to deepen cooperation with Argentina, Brazil, Peru and Chile…[and we] note…the positive experience of cooperation in humanitarian and rescue operations and assistance to countries affected by natural disasters.[33]

In a multi or pluri-polar world order, traditional notions of security become reconfigured under a globalist lens of engagement, and perceptions and applications of security are complex, based on global as opposed to power-politic and state-centric notions of security—which fosters and legitimates so-called unipolar conceptions of "global" security based on power capacity and projection.

> The anarchic quality of the strong nation system and the aggressive, self-maximizing behavior of the strong nation it generates serve to diminish the possibility of achieving [Society. Yet,] the justification for aggressive, self-

DOI: 10.1057/9781137308139

maximizing behavior within the strong nation system is weakened once in the complexity phase, and such behavior and its justification are no longer viable when a related conflict exists or emerges between global security and national security. The signature characteristic of global governance for global problems is that global security takes precedence over national security when a conflict arises.[34]

The idea of a multi or pluri-polar world order provides fecund soil for the growth and importance of soft power and other non-material (intangible) forms of power that are directly and indirectly linked to a society of states framework for world order.

As far as the emergence of complex societal concepts to premise, define, guide, and regulate international relations, Russia's post–Cold War engagement is thoroughly steeped in CCNs; a "new," complex, and viable dimension of power is rapidly emerging on the world stage and effectuating a CAS-based interfacing between and among states. According to Foreign Minister Lavrov,

> [o]ur multilateral cooperation within APEC opens up new prospects. During the Russian chairmanship in 2012 we are going to accentuate the practically significant joint initiatives, including the development of trade and economic infrastructure, cooperation for the modernization of our economies and ensuring energy security...augmenting cooperation, which we intend to build on a pragmatic, de-ideologized, equal and mutually advantageous basis, serves our common interests...and [is] consistent pursuit will bear witness to the fact that a new stage is being established in our relations.[35]

In 2010, Prime Minister Putin and Danish Prime Minister Rasmussen inaugurated the first transatlantic cargo line between Russia and Ecuador in St. Petersburg, which is designed to promote Russian trade with South America. The new line will open up new markets for Russian export, which may provide equipment to the coal-mining areas of Chile, and various other goods, equipment, and services to the agricultural regions of Ecuador, Colombia, and other Central American and Caribbean countries while expanding commercial possibilities for Latin American fresh fruit exports to Russia.[36] Hence "Russia intends to develop, in every possible way, cooperation with countries of Latin America and the Caribbean: 'For a long time Russia has been absent in Latin America and countries of the Caribbean. Now a strategic decision has been made—we will actively develop cooperation...such cooperation 'is very important

DOI: 10.1057/9781137308139

for the balance of forces in the world.' "[37] Evidently, Prime Minister Putin's words quoted within this newsreport carry a revealing and important signal of Russia's present and future relations with Latin America. Its strategic intentions in the western hemisphere are "to balance" strategically and via the use of soft power and CCNs the disproportionate US influence in the region.

Notes

1   Luis Eduardo Guarnizo and Michael Peter Smith, "The Locations of Trans-nationalism," *Comparative Urban and Community Research*, no date, p. 2, accessed on 19 January 2011 <hcd.ucdavis.edu/faculty/.../smith/.../ Locations_of_transnationalism.pdf>.

2   Guarnizo and Smith, "The Locations of Trans-nationalism," p. 3.

3   Sergey Lavrov, "The New Stage of Development of Russian-Latin American Relations," *Ministry of Foreign Affairs of the Russian Federation*, August 24, 2011, accessed on 22 July 2012 <http://www.mid.ru/brp_4.nsf/o/ A27D6F235094016DC32578F70042C31C>.

4   Timothy Bancroft-Hinchey, "Russia Boosts Relations with Latin America," *Pravda.RU*, April 5, 2010, accessed on 20 January 2011 <http://english.pravda. ru/world/americas/05–04–2010/112853-russia_latin_america-o/#>.

5   Stephen Blank, "Russia in Latin America: Geopolitical Games in the U.S.'s Neighborhood," Paper prepared for IFRI Russia/NIS Center, April 2009, accessed on 25 March 2010, www.ifri.org/ ... /ifri_Blank_Russia_and_ LatinAmerica_ENG_April_09.pdf.

6   For examples, see Alex Sanchez, "A COHA Report: Russia Returns to Latin America, Council on Hemispheric Affairs," 2009, accessed on 27 March 2010 <http://www.printfriendly.com/print?url=http%3A%2F%2Fwww.coha. org%2Frussia-returns-to-latin america%2F&partner=sociable>.

7   Nil Nikandrov, "Russia – Latin America: The Union of Solidarity and Pragmatism," *Ria-novosti*, June 21, 2010, accessed on 23 July 2012 < http:// en.rian.ru/international_affairs/20100621/159513144.html>.

8   See Jorge G. Castañeda, "Latin America's Left Turn," *Foreign Affairs*, May/ June 2006, accessed on 1 April 2010 <http://www.foreignaffairs.com/ articles/61702/jorge-g-castaneda/latin-americas-left-turn>.

9   Bancroft-Hinchey , "Russia Boosts Relations with Latin America".

10 Pravda.Ru, English Translation, "Guatemala Wants Russian Arms in Exchange for Coffee and Sugar," March 29, 2010, accessed on 29 March 2010<http://english.pravda.ru/world/americas/23–03–2010/112681-guatemala-o>.

DOI: 10.1057/9781137308139

11   See Marie Mendras, "Soviet Policy Toward the Third World," *Proceedings of the Academy of Political Science, Soviet Foreign Policy*, Vol. 36, No. 4, (1987), pp. 164–175, and, particularly, 169.

12   Tumgazeteler, "A Look at Russian-Latin American Relations as 'New Cold War' Talks Gain Momentum," *Tumgazeteler*, September 26, 2008, accessed on 10 February 2010 <http://www.tumgazeteler.com/?a=4157397>.

13   US Library of Congress, "Country Study: Soviet Union," *U.S. Library of Congress, Federal Research Division*, May 1989, accessed on 18 February 2010 <http://lcweb2.loc.gov/cgi-bin/query/r?frd/cstdy:@field%28DOCID+su0269%29>.

14   Ibid.

15   Ibid.

16   See Edward González, "Institutionalization, Political Elites, and Foreign Policies;" Cole Blasier, "The Soviet Union in Cuban-American Conflict;" and Jorge I. Domínguez, "The Armed Forces and Foreign Relations," both in Cole Blasier and Carmelo Mesa-Lago, (eds), *Cuba in the World* (Pittsburgh: University of Pittsburgh Press, 1979) pp. 3, 37–39, 46–47, and 59.

17   See Mendras, pp. 164–175. For example, Cuba was alleged to have provided Soviet aid packages to Grenada (1979–1983), and Nicaragua and Cuba were alleged to have provided extensive Soviet aid packages to armed opposition groups in El Salvador.

18   US Library of Congress, "Country Study: Soviet Union".

19   Ministry of Foreign Affairs of the Russian Federation, "About the Official Visit Minister of Foreign Affairs of Peru R. Roncayolo to Russia," May 29, 2012, accessed on 31 July 2012 <http://www.mid.ru/bdomp/brp_4.nsf/e78a48 070f128a7b43256999005bcbb3/33566c5d54c747e444257a1600384990!OpenD ocument>.

20   US Library of Congress, "Country Study: Soviet Union".

21   Alex Sánchez, "Russia Returns to Latin America," *Panama News*, February/March 2007, accessed on 4 February 2010 <http://www.thepanamanews.com/pn/v_13/issue_04/opinion_05.html>.

22   Nil Nikandrov, "Russia – Latin America: The Union of Solidarity and Pragmatism".

23   See Mervyn Bain, *Russian-Cuban Relations Since 1992: Continuing Camaraderie in a Post-Soviet World* (Lanham: Lexington, 2008) pp. 129–130.

24   Stephen Blank, "Russia in Latin America: Geopolitical Games in the U.S.'s Neighborhood," pp. 8–9.

25   Vladimir Putin, speech, 43rd Munich Security Conference, 10 Feb. 2007, *BBC News Online*, accessed on 1 June 2007 <http://news.bbc.co.uk/2/hi/europe/6349287.stm>.

26   Sánchez, "A COHA Report: Russia Returns to Latin America, Council on Hemispheric Affairs".

DOI: 10.1057/9781137308139

27   Blank, Russia in Latin America: Geopolitical Games in the U.S.'s
      Neighborhood," p. 8.

28   Guarnizo and Smith, "The Locations of Trans-nationalism,", p. 5.

29   Minister Lavrov quoted in Sanchez, "A COHA Report: Russia Returns to
      Latin America, Council on Hemispheric Affairs"; Alex Nicholson and Andre
      Soliani, "Russia, Brazil Plan to Buy $20 Billion IMF Bonds," June 10, 2009,
      accessed on 1 April 2010 <http://www.bloomberg.com/apps/news?pid=20670
      001&sid=a5nc3eTSovTc>.

30   Ibid.

31   Sergey Kislyak quoted in Alex Sanchez, "Russia Returns to Latin America."

32   DiploNews, "Russian-Cuban relations Are 110 Years Strong," July 20, 2012,
      accessed on 31 July 2012 <http://www.diplonews.com/articles/2012/20120720_
      RussiaCuba.php>.

33   Sergey Lavrov, "The New Stage of Development of Russian-Latin American
      Relations."

34   Bennett Stark, *A Case Study of Complex Adaptive Systems Theory Sustainable
      Global Governance: The Singular Challenge of the Twenty-first Century*, July
      2009, University of Ljubljana, WISDOM RISC-Research Paper No. 5, p. 5,
      accessed on 25 January 2011 <www.wisdom.at/Publikation/.../RRR_BStark_
      SustainableGlobal_09.pdf>.

35   Sergey Lavrov, "The New Stage of Development of Russian-Latin American
      Relations".

36   See "Economía Rusia y Dinamarca inauguran línea transatlántica entre Rusia
      y Ecuador" ("Russia Opens Transatlantic Sea Link With South America"),
      El Universo.com, Guyaquil, Ecuador, March 22, 2010, accessed on 26 January
      2011 <http://www.eluniverso.com/2010/03/22/1/1356/rusia-dinamarca-
      inauguran-linea-transatlantica-carga-rusia-ecuador.html>. *Author's
      translation from the Spanish.

37   Zeenews.com, "Russia Seeks to Boost Ties with Latin America," *Zeenews.
      com*, accessed on 10 February 2010 <http://www.zeenews.com/printstory.
      aspx?id=601165>.

DOI: 10.1057/9781137308139

4

Building Complexity: Select Case Studies of CCNs— Russia and Latin America

Abstract: *A select analysis of Russia's engagement with Mexico, Colombia, Venezuela, Cuba, Nicaragua, Bolivia, Brazil, and Ecuador is conducted to identify how complex cooperative networks facilitate society. The case studies are not meant to provide an exhaustive and comprehensive analysis, but to provide initial and suggestive examples of how networks are affecting statecraft and international organization in an emergent society of states. Overall, it seems that networks are becoming a viable and effective means of facilitating interstate relations. Networks are creating viable alternative avenues and forums for productive and mutually beneficial engagement. Commercial relations, e.g., trade in fruits, meats, flowers, and oil, and humanitarian aid, are among the types of networks laying the foundation for more intensive and extensive cooperative partnerships.*

Astrada, Marvin L. and Martín, Félix E. *Russia and Latin America: From Nation-State to Society of States.* New York: Palgrave Macmillan, 2013. DOI: 10.1057/9781137308139.

DOI: 10.1057/9781137308139

At the outset, it must be clarified that the following case studies are employed in a stylized and selective manner to initiate a process of identifying and empirically mapping how Complex Cooperative Networks (CCNs) may be impacting Russo-Latin American relations. In particular, emphasis is placed on the military, trade, investment, energy, security, cultural exchange, and resource exploitation dimensions of these countries' engagement via CCNs. The aim of this chapter is to begin the process of identifying how complex cooperative partnerships facilitate and are facilitated by CCNs. Accordingly, the case studies are not meant to provide an exhaustive and comprehensive analysis of all facets of Russia–Latin America relations, but rather to provide suggestive examples of how CCNs are affecting statecraft and international organization in an emergent society of states. As explained in the first two chapters of this study, the approach developed is an attempt to think outside the well-established traditional parameters of explaining and understanding international order and organization. This is work thus seeks to establish a working context for analysis. Given the theoretical nature and scope of this work, the empirical evidence employed is to chart and illustrate an alternative explanatory framework for causal sequences in world politics. Therefore, the empirical evidence presented is based mainly on data figures, and political and economic analyses culled from various public information and secondary sources. We, then, examine these data in light of the theoretical argument comprised in the first two chapters of this study in order to determine if and how the empirical evidence, in fact, bolsters the contention that emergent CCNs are, indeed, assuming a transformative role in world politics. The purpose of this work is to identify the nascent and discrete development of CCNs and how they are impacting interstate relations, with the realization that, though challenging a system of states framework, CCNs have not displaced nation-states or the principles of *realpolitik* in the conduct of world politics. Based on the assumption that CCNs are a relatively recent phenomenon, thus, the tentative and limited scope of the empirical evidence employed must be keep in mind. In the rest of this chapter we present various sections on specific Latin American countries and their ongoing relations with Russia. We begin with an analysis of Russo-Mexican relations, followed by similar analyses of Colombia, Venezuela, Cuba, Nicaragua, Bolivia, Brazil, and Ecuador. The order of these cases follows no particular criteria other than a

DOI: 10.1057/9781137308139

desire to demonstrate, using the best available evidence, the increasing relevance and importance of CCNs in Russo-Latin American relations and, generally, in world politics.

Mexico

In the case of Russia–Mexico relations, Mexico has shown receptiveness to engaging in a variety of partnerships with Russia. Engagement has been premised on diversified trade and investment agreements due, in part, to the fact that it requires a massive amount of support—logistical, technical, financial, diplomatic, political, and military—in its ongoing struggle to rein in the illegal narcotics trade that plagues the government's stability, legitimacy, and authority. While not limited to combating narco-trafficking, Mexico's need for resources, contacts, and networks to deal with the problem has made it amenable to engaging Russia. The social, political, and economic context that presently defines Mexico makes it open to receipt of foreign assistance and investment. Indeed, the illicit narcotics trade tinctures Mexico's international relations; Mexico remains one the most significant sources for the flow of illicit drugs in the Americas. The level of violence that accompanies the drug trade has risen to mammoth proportions, presenting Mexico with profound challenges to its authority, capacity to govern, and legitimacy. Mexico has, thus, accepted Russian promises to aid it in its struggle to combat the illegal drug trade.[1]

While Mexico welcomes Russian assistance for counter-narcotics efforts, trade and investment, as well as cultural exchange, are also viewed as being important and beneficial for its short and long-term wellbeing. Trade and partnerships have occurred on a variety of levels. The Secretariat of Foreign Affairs of Mexico, e.g., notes that Russian investment reached $ 1.2 million between 1999 and 2010, allocated to the service sector (70%), commercial sector (20%) and the manufacturing industry (10%), with 47 enterprises operating with Russian capital and in August 2010 the Secretariat of Energy of Mexico has signed a cooperative declaration of intent to develop nuclear energy for civilian purposes with Russia's Rosatom.[2] In January 2011 the Foreign Affairs Office disclosed joint actions being taken by both countries in the realms of cultural, scientific, technology, investment, and economic cooperative relations between Mexico and Russia.

DOI: 10.1057/9781137308139

Russia–Mexico relations are thus becoming a conduit for complex channels of cooperative engagement. Both countries have held continual high-level political talks regarding the establishment of major energy partnerships since 2003.[3] Tourism has formed the basis for CCN to facilitate society between Russia and the region. Russia and Mexico have officially arranged for there to be direct flights between Moscow and Cancun, facilitating the exchange of tourists between the countries.[4] In addition to tourism, Mexico's second largest airline, Interjet, has signed a "$650 million deal with Superjet International to buy 15 regional Sukhoi Superjet 100 planes. This is the first delivery contract for Russia's new type of Superjet airplanes to a Latin American country...deliveries of Superjet airplanes are expected in the second half of 2012."[5] Russian tourism businesses also view Mexico's regional trading partners in other parts of Central America as viable places for investment, for establishing CCNs. The Salvadoran Minister of Tourism has observed that

> part of the Russian market is in a strong economic momentum, this coupled with demand for exotic destinations, makes Central America one of the sites selected when they think about traveling. Therefore we are betting on long-range markets that will be strengthened through a new Iberia air link, starting on October 2010.[6]

Tourism is but one CCN that has begun fostering society between Mexico and Russia. Overall, Mexico stands to benefit far more from its relations with Russia—a benefit that is not limited to military and security interests. Recent high-level interaction between Mexico and Russia point to a viable basis for the substantive expansion of contacts and engagement that transcend traditional and limited bases for engagement—and it is interesting to note how multilateralism remains in the fore and background of Mexico–Russia relations. For instance, during October 7–9, 2008, the Mexican Secretary of Foreign Affairs officially visited Russia. During meetings with the Russian Minister of Foreign Affairs and Minister Economic Development, Russia and Mexico formally declared their support for a multipolar world order, and expressed their desire to cooperate closely in a variety of joint endeavors such as counterterrorism and combating organized crime. "Noting the positive dynamics of Russian-Mexican relations, the sides expressed their reciprocal readiness to expand mutually advantageous cooperation in the economic, commercial, scientific, technological, cultural and humanitarian fields.

DOI: 10.1057/9781137308139

Emphasis was laid upon the importance of intensifying the mechanisms of business cooperation."[7]

In the realm of economic and commercial relations, Mexico has developed intensive and extensive cooperative relationships with Russia. According to the Russian Embassy in Mexico, during 2009, trade and economic relations between Russia and Mexico developed on the basis of growth of both economies. According to data provided by the Mexican Ministry of Economy,[8] bilateral trade between Russia and Mexico reached $609.64 million in 2006. In 2009 the difference between Russian export and Mexican import decreased substantially totaling $289.98 million against $490.90 in 2008. The decrease by 15.94% of Russian export was fixed, and this indicator fell to $449.81 million against $535.09 million in 2008. Concurrently, Mexican import increased by 361.66% and reached a record mark that amounted to $159.83 million versus $44.19 million in 2006.[9]

In comparison with 2006, the present structure of Russian export has been intensified and diversified. According to the Russian Embassy in Mexico, Russian export consisted of chemical products such as fertilizers totaling approximately $236.52 million (55.3%). Of this category of export, of the most significant was carbamide (32.8%), nitrate and phosphate fertilizers (12.3%), combined fertilizers (5.4%), ammonium saltpeter (2.7%), caoutchouc (1.4%), and kalium fertilizers (0.7%). Other chemical product exports consisted of semi-finished products of ferrous and non-ferrous metals ($96.31 million) and machines and equipment ($70.68 million).

The substantial growth of Mexican import and export vis-à-vis Russia reflects a growing and complex trade and economic dimension lacking under the USSR approach to the Americas. Deep commercial and investment ties have resulted in the instauration of a viable, complex, cooperative, and non-militarized basis for intestate relations that, despite traditional strategic interest that may be at play, are providing the potential for engagement on a societal level. Travel, tourism, and trade provide linkages that foster connectivity. Other types of bilateral trade include Mexican automobiles and quadrocicles ($120.1 million), seamless steel pipes and profiles ($2.36 million), and freezing equipment ($1.79 million.[10] In addition to such products, Mexico has stepped up its export of foodstuffs, cultural products, and liquors to Russia—for instance, Mexican export of tequila ($11.49 million), beer ($3.01 million), toys ($2.41 million) and deodorants ($1.74).

DOI: 10.1057/9781137308139

Since 2005, Mexico and Russia have sought to further deepen integrative cooperative networks based on trade and mutual investment via CCNs of a variegated nature. At the present time, Russia–Mexico relations have continued to intensify. Several trade and investment agreements have been signed in the last few years which have the capacity to further intensify and deepen financial and commercial links. In 2004, e.g., the Mexican bank Bancomext signed a series of agreements with Russian banks, and in 2005 agreements on aerial standards, health-related cooperation, and a letter of intention for cooperation in the field of energy were signed.[11] Additionally, a convention to avoid double income-tax taxation became effective in 2009 and a business consortium, the Mexico-Russia Corporate Group, held its first meeting in 2007.[12] Given the increasingly integrated relations between Russia and Mexico, some commentators have speculated that intensified bilateral economic cooperative networks in electro-energy, technical-military and space exploration may be a possibility in the near future.[13]

Colombia

Colombia presents a hybrid case, in which Russia has found a receptive context in militarized and non-militarized bases for interstate relations. Military-security interests of Colombia encompass its concern over what it perceives to be the militarization of the Venezuelan state and its long-term problem of domestic terrorism, and its non-militarized interest to diversify trade relations with foreign powers besides the US. CCNs have arisen, but *realpolitik* considerations continue to play a prominent role in Colombia's engagement with Russia.

Relations between Venezuela and Colombia have been steadily deteriorating since Hugo Chávez took power in 1998. Hard power considerations and interests revolve around territorial integrity and political stability, and soft-power interest and considerations are primarily economic in nature, revolving around commercial relations and financial investment. Colombia seeks to fortify itself against Venezuela's militarization as well as eliminate the threat of the The Revolutionary Armed Forces of Colombia—People's Army (Spanish: *Fuerzas Armadas Revolucionarias de Colombia*—Ejército del Pueblo (FARC-EP)) guerrillas through hard and soft power. In the case of the FARC, Colombia has been determined to use its security forces to maintain pressure that

DOI: 10.1057/9781137308139

has caused the FARC to lose control of physical territory and degraded FARC command and control capacity. Among the major successes of the Colombian military in 2008 were the deaths of key FARC leaders, including members of the ruling Secretariat, a continued high number of FARC desertions, and the 2 July rescue of 15 hostages, including three US citizens.[14] Despite these reverses, the FARC remains a viable security threat based on profits obtained from narco-trafficking and cross-border sanctuaries in the territory of Venezuela and Ecuador.[15]

In addition to providing havens, Venezuela is accused of providing military hardware in addition to political support for FARC insurgents. Both the FARC and Venezuela have prompted Colombia to augment its military capacities. Colombia has thus sought to acquire military ordnance from Russia. In October 2008 the Colombian Defense Minister made an unprecedented visit to Russia to discuss and formalize cooperative military relations, and to attend an Interpol police conference. The Defense Minster's official visit was very significant for Russia–Colombia relations. This is the case because high-level political engagement took place for the first time, and in addition to military-security concerns, engagement focused on establishing cooperative networks to combat narco-trafficking and terrorism as well as discussing a new defense accord and negotiating to purchase helicopters and radar systems.[16]

Although Colombia has expressed serious concerns over the flow of Russian arms to Venezuela, it is interesting to note that Colombia–Russian relations have been in the process of substantive development and diversification since the early 2000s, with the theme of multilateralism having been explicitly stated in official descriptions of Colombian-Russian shared interests. Colombia has stated that, at present, Russia and Colombia have similar positions in light of the major problems affecting international relations. In particular, they each consider the installation of a multipolar world to be in the best interest of both countries and international order in general. Colombia and Russia agree that an international al order premised on multilateralism (what Russia and some countries in the region have referred to as pluri or multi-polarity) as opposed to uni-polarity will better ensure that "there is no room for arrogance or the power of money, and that a redistribution of wealth for the benefit of sustainable development of all nations, especially the poorest, is desirable as well as obtainable" in such an order.[17]

DOI: 10.1057/9781137308139

Some significant developments in Colombia–Russia relations that have laid the foundation for developing substantive soft-power networks via CCN in the realm of economics and commercial activity involve a visit to Colombia of an official delegation of the Russian Comptroller General to attend a meeting of the General Comptrollers of Europe and America during July 10–11, 2002. The meeting was devoted to exploring multilateral (cooperative) ways of addressing the problems of corruption, tax evasion, money laundering, and other forms of financial crime.[18] In the field of juridical cooperation, official meetings were held in Bogota (October 2002) between representatives of both countries to discuss establishing cooperative networks between Russia and Colombia in international criminal matters. In discussions pertaining to combating narco-trafficking, domestically and abroad, consultations were held in Bogota (November 2002). The consultations were aimed at encouraging the constructive exchange of experiences so as to develop ways to coordinate each state's efforts in addressing the problem.[19]

In the realm of complex networks based on soft-power interest and considerations, Colombia has viewed Russian engagement as beneficial for Colombia's short and long-term economic interests and development. The IV Meeting of the Colombian-Russian Intergovernmental Commission on Economic, Trade, Scientific and Technological Cooperation (April 2009), held in Bogota, to reconsider a broad range of topics, such as "trade, investment, customs, energy and fuel cooperation, science and technology, as well as cooperation in the sphere of army technology, telecommunications, and the judicial and consular field."[20] Russia and Colombia took the opportunity to discuss a diversified approach to address common problems and interests via CCNs. Proposals were based on each respective country's "interests and knowledge of the possible spheres of joint actions, and these were left for a feasibility analysis."[21] Other fields of cooperative engagement that have been considered and/ or developed by Colombia and Russia involve extensive and intensive commercial exchange agreements, e.g., in the sale of flowers, coffee, cacao, textiles, beef, chicken, fish and shrimp, and in travel and tourism, e.g., for visa-free travel between the respective countries.[22]

Thus, while *realpolitik* concerns continue to underlie Colombia–Russia relations, nevertheless, nascent CCNs are being developed between the two countries and the prospects for cooperation, beyond a purely military-security basis, have the potential to develop in the near future.

DOI: 10.1057/9781137308139

Venezuela

In the case of Venezuela, as with Colombia, military and non-military considerations are both at play in interstate engagement. Venezuela has proven to be one of the largest importers of Russian ordnance since Cuba during the Cold War. Yet, substantive soft-power networks are also being utilized in tandem with *realpolitik*. From Russia's vantage point, Venezuela can be considered "a strategic partner, very rich in natural resources and with great opportunities in particular, regarding the exploitation of petroleum, gas, aluminum and gold reservoirs, as well as the construction of railroads, underground transportation networks, hydraulic works and the transfer of modern technology."[23] Within a *realpolitik* context, in which Russia has assumed the role of arms supplier, it has also engaged Venezuela on a variety of non-military levels, such as, energy partnerships, military exercises, technology transfer agreements, and development of nuclear energy capacity. Russia's role as arms provider should not be construed as an indication that CCNs have no place or do not have an independent effect on interstate relations. Indeed, the

> two countries already co-operate closely on energy matters, with their state-owned energy companies embarking on joint enterprises [...and] Latin American governments of all political colors have themselves in recent years been seeking new commercial and diplomatic allies in what they see as a changing, multi-polar world.[24]

As with Mexico and Colombia, multipolarity (multilateralism being the less politically charged term) is a driver of engagement and relations with Russia. Multipolarity is in line with societal notions of engagement on a variety levels via CCNs. Efforts to establish substantive non-military relations have taken place in the realm of energy development and economic cooperation between Russia and Venezuela. Economic relations are being cultivated within new as well as preexisting legal frameworks established during the Soviet era. In 2003, e.g., collaborative statements to implement anti-dumping policies were signed, and both countries sought to establish an agreement to address issues of taxation and financial crimes, and in 2007 several agreements were reached in the fields of agriculture, energy, industry, technology and trade.[25]

Cooperative networks in the energy sector have proliferated. The establishment of CCNs has resulted in Russia and Venezuela using and

DOI: 10.1057/9781137308139

building upon a non-militarized basis upon which to engage one another for mutual self-interest.

> Russian companies Gazprom and Lukoil have signed agreements with Venezuelan state oil company Petroleos de Venezuela SA to jointly explore several Orinoco fields [and to invest approximately $30 billion in developing refinement and exporting capacities]...five of Russia's biggest oil companies are looking to form a consortium to increase Latin American operations. State-controlled Rosneft, Lukoil, Gazprom Neft, Surgutneftegaz, and TNK-BP hope to build a $6.5 billion refinery to process Venezuela's tar-like heavy crude."[26]

In February 2010, Russia and Venezuela signed an official agreement to extract crude oil from the Orinoco region,[27] and in October 2010, Russia and Venezuela signed the 2014 Strategic Plan of Action for the Development of Relations between Russia and Venezuela. The establishment of such agreements began to lay the foundation for CCNs to grow and proliferate. Cooperation has extended to the development of nuclear energy for peaceful purposes, as well as agreements to engage in joint technology development and transfer in the realms of agriculture and manufacturing. In 2010, Venezuela secured a significant nuclear energy and space technology agreement with Russia, wherein Russia has agreed to aid Venezuela in the construction of a nuclear power plant and in developing a space industry. Venezuela has made it a point to emphasize how the recent agreement is exemplary of the multipolar enterprise (that it shares in common with Russia) that it seeks to effectuate, and that the agreement is indicative of Venezuela's ties with greater powers that share common interests in energy, military ordnance, and the structure of international relations.[28]

Cuba

Cuba has found itself in the throes of economic and ideological stagnation since the end of the Cold War. With Fidel Castro stepping down from power in July 2006 due to health reasons, Cuba has struggled to maintain its political, ideological, and economic integrity. The military has assumed a cardinal role in maintaining the regime's power under the rule of Raúl Castro. Within a context pervaded by economic stagnation and the exacerbation of social and political stability, Cuba has and

DOI: 10.1057/9781137308139

continues to welcome aid from foreign powers to sustain itself. In an effort to see its way through the profound economic challenges it faces in the present international order, where it continues to be found not in compliance with emergent societal norms of human rights and the dignity of the human being as defined by various international principles, customs, traditions, and formal conventions (e.g., The UN Declaration of Human Rights), Cuba seems to have embraced CCNs in its relations with Russia to revitalize its stagnant economy.

Russian-Cuban relations suffered a significant downturn in the years following the demise of the USSR. Trade between 2004 and 2008 was well below the levels during the Soviet era. Presently, Russia and Cuba have shown a growing interest in strengthening bilateral collaboration in a variety of spheres. Within a post–Cold War context, Russia continues to be a major creditor for Cuba, and both countries maintain close economic ties. Since 2008, Cuba and Russia have significantly increased cooperative ventures with each other via trade, investment, and natural disaster relief. For example, Russia was the first country to provide aid to Cuba after it experienced multiple devastating hurricanes (2008) in the form of food donations, medical supplies, and construction supplies.[29] In November 2008, then-Russian President Medvedev visited Cuba to strengthen economic ties and to allow Russian companies to drill for oil offshore in Cuban waters and mine nickél in Cuba.[30] Raúl Castro traveled for a weeklong visit to Moscow from January 28, 2008 to February 4, 2009; the talks included $20 million worth of credit to Havana, and 25,000 tons of grain as humanitarian aid to Cuba.[31] In July 2009 Russia began oil exploration in the Gulf of Mexico after signing a deal with Cuba. Under the new agreement, Russia has also granted a loan of $150 million to buy construction and agricultural equipment.[32]

Russia is thus not limiting itself to the most powerful or the most weak or most anti-American countries—all countries that wish to open up trade, diplomatic, and communication networks are being engaged by Russia. Russia, along with other countries such as China, has provided Cuba with crucial economic sustenance in the present international context. According to the US Dept of State (March 29, 2010), Cuban trade in the form of export markets to Russia in 2008 was $71 million, and has spent $269 million on Russian imports.[33] Despite the difficulties that both countries have experienced transitioning from the former Soviet mode of engagement, i.e., one premised on economic subsidization on the basis of ideology and Cold War and (bipolar) superpower politics, to the present

DOI: 10.1057/9781137308139

international order, Russia and Cuba have demonstrated a willingness to reengage and repurpose relations via CCNs. In 2006, e.g., Russia signaled

> a new effort to expand trade and investment, albeit financed by Russian credit. Russia set aside, for the moment, more than U.S. $20 billion in Soviet-era debt, re-structured post-1991 debt, and extended a new credit line to Cuba. The new credit line is for U.S. $355 million repayable over 10 years at an interest rate of 5%. The new credit is conditioned in that it must be used to purchase Russian cars, trucks, and planes, as well as to finance Cuban energy and transport infrastructure projects, including air navigation systems. Russia further agreed to restructure U.S. $166 million in debt accumulated since 1993...Raúl Castro made a state visit to Russia in February 2009 during which several additional trade agreements were signed.[34]

Despite the loss of the very close ideological, military, and security partnerships that defined Cuba–Soviet relations for the duration of the Cold War, Russia has reengaged Cuba via CCNs. The diversification of trade, the move away from politicized and ideological discourse as a primary basis for engagement, and the use of legal frameworks to buttress, develop, and enhance trade and investment point to a profound shift in interstate relations between Russia and Cuba. Russia and Cuba are moving toward reestablishing diversified and substantive ties that transcend ideology and the old superpower politics.

> In July [2008] Russian Security Council Secretary Nikolai Patrushev and Deputy Premier Igor Sechin led a large delegation to Havana to meet with both defense and economic officials. Following the visit, the Russian Security Council issued a statement announcing that Russia and Cuba planned 'consistent work to restore traditional relations in all areas of cooperation'."[35]

In sum, Russia–Cuba relations have become more cooperative and diversified. Both countries have signed agreements that seek to facilitate, strengthen, and enhance cooperative networks. Although Cuba resides in an ever-present past, it has realized that CCNs are invaluable in providing life-sustaining capital, investments, and interactivity in a networked world.

Nicaragua

Russia's position toward Nicaragua was made very clear in 2008: "Russia would like to have fully-fledged long term relations with Latin America,

DOI: 10.1057/9781137308139

and, in particular, with such an important partner as Nicaragua."[36] During high-level meetings between Russian and Nicaraguan officials, Nicaraguan President Daniel Ortega stated that "Multipolarity [i.e., multilateralism] is a reality...Extreme conditions are being created in Latin America and all the governments are welcoming Russia's presence. We are ready to develop relations in all spheres that will create prosperity and progress for our peoples."[37] Russia has approached Nicaragua as it has the region, i.e., seeking to establish integrative CCNs to realize strategic as well as non-military cooperative interests. CCNs have the potential to "enhance the expansion of the Russian economic presence in Latin America... [based on] participation in major economic projects in Latin America [and] expansion of trade relations with...fully meets Russia's national interests."[38]

Russia and Nicaragua have signed or are in negotiations to sign cooperative agreements that cover a range of accords on space technology, global positioning systems, constructing an alternate transoceanic route to the Panama Canal, and energy investment and development. Nicaragua has welcomed Russian engagement for political and economic interests, as well as the shared goal of establishing a multilateral international order. Ideological predilections aside, there seems to be a growing realization in the region of the value of CCNs in creating integrative non-militarized partnerships based on developing the full potential of states to participate in and benefit from societal notions of peace, prosperity, cooperation, and mutual benefits derived from non-militarized bases of interactivity vis-à-vis interstate engagement. International relations do not have to be solely concerned with security and *realpolitik* concerns to the detriment of other equally important values and interest. North Korea, for instance, is an example of a state wholly devoted to *realpolitik* at the dire expense of other important interests—namely, economic prosperity and high material quality of life for its citizens.

Russia–Nicaragua relations, in spite of lingering ideological rhetoric on the part of Nicaragua, seem to have transcended the myopic focus on ideological affinity between the countries. As in the case of Cuba, it appears that Russia views ideology as a non-sustainable and unproductive form of interstate engagement. Russia as well as Cuba and Nicaragua have "learned" that a more efficient and effective means exists for the conduct of interstate relations, i.e., CCNs. Outliers that either cling to a purely system of states view, e.g., North Korea, or insist on retaining high degrees of ideology (e.g., Venezuela) or militarized-securitized bases as

DOI: 10.1057/9781137308139

the primary basis for interstate relations have not "learned" how to function effectively in the present networked international order. In the case of Russia and Nicaragua, each country has committed itself to developing non-*realpolitik* bases for relations. Russia has, for example, offered to provide help to Nicaragua so it can begin to better develop its capacities for oil and gas exploration, and both countries have discussed Russian participation in developing an alternative route linking the Pacific and Atlantic oceans by way of a new canal. Although such plans are in the early stages, and there is the possibility that such projects may not materialize immediately[39] due to global and domestic economic conditions, the act of conceptualizing and laying the foundation for CCN to effectuate such ambitious goals is indicative of a change in the modus operandi and competing societal interests to realpolitik considerations.[40]

In the realm of trade, Russia has signed a directive opening a long-term Russian trade mission in Nicaragua. This act is based on Nicaragua being a major exporter of foodstuffs, in particular coffee, nuts and tobacco, to Russia. Russia provides exports to Nicaragua in the form of machinery, equipment, and chemical products. Furthermore, diversification of relations has been discussed in the realms of oil exploration in the Caribbean and the Pacific Ocean as well as developing and/or refurbishing Nicaragua's construction of roads and bridges.[41] Also, is in the case of Colombia, Russia and Nicaragua have signed an agreement to introduce visa-free travel for their citizens, as well as a memorandum of understanding between Nicaragua's Petronic and a consortium of Russian oil companies on cooperation on oil and energy projects. Lastly, Nicaragua also plans to supply 12,000 metric tons of beef and pork products to Russia annually from 2010 onwards as well as strategize with Russia on how to further diversify its export regime.[42]

In sum, interrelations have taken a serious cooperative, non-militarized turn, and soft-power resources and CCNs are being employed to diversify and integrate the countries. Efforts aimed at providing humanitarian aid to Nicaragua, jointly addressing issues related to combating narco-trafficking, sharing of technology, and engaging in joint energy development partnerships all provide building blocks for developing viable alternative bases for interstate relations based on CCNs. As noted by Dmitry Medvedev, Russian-Nicaraguan relations involve a

> high degree of mutual understanding and the focus on achieving tangible, practical results. In recent years much has been done to enhance political, economic and humanitarian ties. Together we have successfully

DOI: 10.1057/9781137308139

strengthened emergency response cooperation... [Russia is] determined to continue active cooperation with Nicaragua in international affairs. The prevailing practice of coordinating positions allows us to promote a common approach in the interests of our nations, in guaranteeing security and sustainable development of a multipolar world.[43]

Bolivia

In the case of Bolivia, the core of relations with Russia is mainly economic as opposed to political, although acquiring Russian military ordnance remains high on the list of Bolivia's interests and priorities. Overall, Russia–Bolivia relations are premised on the development of bilateral cooperation in the realms of investment, trade, and energy partnerships. Economic development in the form of agreements with Russia expands Bolivia's (and the region's) economic opportunities for development by expanding and diversifying relations.[44] Diversification and development of relations beyond traditional *realpolitik* through CCNs enables both Russia and Bolivia to strengthen rather than weaken the overall health and prosperity of the respective polities. Trade and investment, cultural exchange, agreements based on the rule of law, suppression conventions and humanitarian aid, among other types of CCNs, enable both countries to hook into and participate in a globalized and evolving international order. While security and military concerns remain viable and indispensable, they are not or rather should not be the alpha and omega of interstate relations.

Economic development in the form of CCNs, e.g., institutionalized as well as singular cooperative trade and investment agreements provide opportunities for individual and collective development that in turn, provide a stable basis for additional cooperative partnerships that produce other CCNs which further strengthen cooperative ventures as a more efficient modality of engagement. In September 2008, Gazprom signed an agreement worth $4.5 billion to explore for gas in Bolivia, and in December 2008 Russia invested $4 million in a study on the Bolivian gas industry, with hopes of opening a joint Russian-Bolivian center on gas exploration beginning in 2009.[45] In March 2009, Russia and Bolivia signed a protocol agreement aimed at strengthening democracy in each nation, and in February 2009 President Evo Morales visited Moscow, the first ever visit by a Bolivian head of state. As in the case of the Colombian

DOI: 10.1057/9781137308139

Defense Minster's unprecedented visit, such high-level contacts create the potential for laying down foundations for CCNs to emerge and develop. Russia and Bolivia signed an agreement bolstering energy and military partnerships as well as joint efforts to combat narco-trafficking.[46] In May 2009, Bolivia's Vice Minister of Foreign Affairs said that Bolivia would be making a multimillion dollar arms and transportation purchase from Russia in efforts to combat drug smuggling and production in Bolivia,[47]and in October of that same year Morales announced plans to construct a technical support and repair facility for Russian aircraft in Bolivia.[48]

Interestingly, though, CCNs and non-militarized bases of engagement have also created opportunities for transnational criminals to take advantage of the open and cooperative nature of some aspects of interstate relations. Gambling, e.g., has proven to be a lucrative industry imported from Russia into Bolivia, perhaps suggesting that there are downsides to the hyper-networked nature of present international organization. Gambling has ushered in Russian networks of transnational organized crime and corruption in tandem with legitimate gambling businesses. Russia has banned casinos in most of its national territory, confining the multibillion dollar gambling industry to Siberia;

> [s]ince then, gambling operations have been trying to win back lost profits in [Latin America.] "Bolivia is a gambling company's paradise."...In the last year...gambling operations here have nearly doubled—there are now more than 80 casinos and about 10,000 gambling machines in his country of 9 million people....Once a gambling company is granted a license to operate, there are no limits on the number of sites it can open in Bolivia.[49]

Bolivia's relaxed gambling market has provided lucrative investment opportunities for the legal Russian gambling industry. For example, the Russian company Ritzio International, the largest gambling enterprise in Eastern Europe, "does $1.2 billion in business annually at its 1,000 venues worldwide," and has 15 casinos in Bolivia's major cities.[50] It is important to note that the gambling industry exported from Russia is not confined to Bolivia. Other countries are also becoming sites for gambling operations, such as Mexico, Colombia, Chile, Peru, and Argentina[51]

Regarding energy relationships, Russia and Bolivia have signed various agreements to enhance and facilitate cooperative partnerships. Both countries have actively engaged in talks to open up gas and oil exploration in Bolivia.[52] Thus, although nascent, CCNs are finding their way

DOI: 10.1057/9781137308139

into the mundane and the profound as far as interstate engagement and relations. With CCNs, avenues for cooperative engagement have opened up creating viable alternatives to premising relations solely on *realpolitik* interests and concerns.

Brazil

In the case of Russia–Brazil relations, Brazil has engaged Russia as it has virtually every other trade partner—with enthusiasm and intensive and extensive engagement that integrates Brazil into the economic (and to a lesser extent, cultural) dimensions of its partners. Brazil, as a regional power that seeks to use integrative networks to their fullest potential, seems to have used CCNs to enhance its soft power and economic power. Diversification of the Brazilian economy has resulted in diverse CCNs, as is evident in the Russia–Brazil relations. High-level political contacts have effectively facilitated cooperation between Russia and Brazil. Russia and Brazil have concluded various agreements in the fields of trade, technology, energy, and research and development in the space industry. The fact that Brazil has a sophisticated, diversified, and developed economy has enabled Russia–Brazil relations to become highly efficient, productive, and beneficial to both countries.

Diversification of engagement via CCNs is readily apparent in Russia–Brazil relations.[53] Russia and Brazil recently signed a "memorandum on nuclear cooperation," including projects focused on uranium exploration. The memo encompasses peaceful uses of nuclear energy. The memo

> calls for the development of uranium prospecting technology and the design of new reactors, as well as the design and construction of nuclear research reactors. Rosatom has proposed assistance with uranium exploration, due to the fact that Rosatom chief Sergej Kirijenko opines that Brazilian uranium resources could be increased by a factor of three or even then, if state-of-the-art exploration techniques would be used.[54]

Russia's interests in Brazil and the region in general revolve around steadily and substantively diversifying and increasing the volume of bilateral trade Russia has with Brazil and the region as a whole. Trade has been projected to reach up to $15 billion in 2008 and beyond, and thus remains a top priority for Russia as well as Brazil, which has global

DOI: 10.1057/9781137308139

power aspirations. Economic partnerships, trade agreements, and joint cooperative ventures in non-military spheres create prospective touchstones for diversified CCNs. Then-Russian President Medvedev's major tour of Latin America began in 2008 "in Peru where he signed a series of economic and political co-operation accords. He then traveled on to Brazil, the region's key economic powerhouse. Russia is keen to press the benefits of Russian technology in the areas of oil exploration, aerospace and defense."[55]

Russia seeks to establish CCNs with Brazil via intensification of political dialogue and extension of economic ties. According to ITAR-TASS, the Russian state news agency, political dialogue encompasses cooperative engagement to address and resolve problems associated with the preservation and protection of democracy and human rights, as well as addressing issues of peace, international stability, and the prevention of international conflicts: Brazil has stated that Russian-Brazilian cooperation in international affairs is indicative of

> the strength of principles and values which our countries share in their attitude to key matters. They include the protection of democratic principles, peace and international stability, disarmament and non-proliferation of weapons of mass destruction, efforts to improve international security, cooperation in the struggle against terrorism and the drug trade, and the strengthening of a multi-polar world.[56]

It is interesting to note, nonetheless, how the countries discussed have all tendered the alternative notion of a multilateral world in which cooperative engagement is held out as the primary means to conduct interstate relations. CCNs are vital to such an ordering of international affairs, and continue to develop as states engage and become linked into the global network that defines the present international order.

Ecuador

In November 2008, the Russian Minister of Foreign Affairs, Sergei Lavrov, made an official visit to Quito in order to strengthen and enhance bilateral trade, cultural exchange, and high-level political contacts. Russia expressed interest in building cooperative networks to strengthen bilateral relations, especially in trade, investment, and cultural exchange. A major agreement for facilitating CCNs was established

DOI: 10.1057/9781137308139

in October 2009, when President Rafael Correa made an official visit to
Russia. The visit was historic because it was an unprecedented one—
i.e., as in the cases of Colombia Bolivia, an important step in fostering
CCNs took place when Correa became the first high-level official, the
first Ecuadorian head of state, to visit Russia. Correa's visit resulted in
both countries signing the

> Intergovernmental Russian-Ecuador Agreement on Cooperation and
> Mutual Aid in Customs Affairs...an extensive document, aimed at the crea-
> tion of the necessary contractual legal framework for a large-scale coopera-
> tion in customs affairs...[that] involves mutual assistance of Russian and
> Ecuador customs services within their competence and investigation and
> restraint of violations of customs legislation under internal laws.[57]

Additionally, in March 2012 both Russia and Ecuador signed the
Protocol of Exchange of Information of Pre-arrival of Goods Traded
between the Republic of Ecuador and the Russian Federation. This
agreement had the effect of clarifying the positions of both countries
to strengthen bilateral relations; "the signing of the Protocol represents
a milestone in the realization of the project that will establish direct
flights to transport perishable products, especially roses from Ecuador
to Russia."[58]

Russia has signed a variety of other agreements ranging from help-
ing Ecuador develop a nuclear energy program for civilian purposes
to improving commercial relations to flower markets. The Ministry of
Electricity and Renewable Energy stated that the Russian State Atomic
Energy Corporation (Rosatom) would provide crucial "support and
assistance" to Ecuador to develop nuclear energy capacity. "The
Agreement will allow the development of joint activities in the inves-
tigation of nuclear technologies that can be applied in Ecuador. 'The
cooperation with Russia cannot be measured in economic terms, but in
the benefit that this technological transfer means to developing countries.'"[59] Such a development would certainly have been inconceivable
within the Cold War context, where ideology and *realpolitik* worked
hand-in-hand to divide countries, sowing the seeds of suspicion,
fear, and a sense of foreboding when dealing with the "other" within
the binary of US/USSR international order. Hence, the emergence of
CCNs, which do the opposite—i.e., foster cooperation, engagement,
integration, and provide alternative bases for productive and constructive interstate relations. The Ecuadorian Ministry of Electricity and

DOI: 10.1057/9781137308139

Renewable Energy signed a "memorandum of understanding" with the Russian State Atomic Energy Corporation (Rosatom) to "carry out research in technologies and nuclear devices...Rosatom would provide 'support and assistance' to help Ecuador draft a set of laws that will limit the use of nuclear energy to peaceful purposes only."[60]

Ecuador has also premised engagement with Russia in the form of developing commercial contacts that encompass a wide array of business relationships. Commercially, Russia has expressed interest in establishing business contacts between Russia and the Port of Manta. Russia's ambassador to Ecuador, Valentín M. Bogomazov, during a visit to Manta, sought to discuss specific projects involving bilateral cooperation between Russia and Ecuador's port of Manta by establishing initial contacts and then fostering cooperation between the port of Manta and Russian cities.[61] Such economic and trade contacts could certainly stimulate bilateral trade, deepening cooperative networks between the two countries. Indeed, the ambassador noted that Russia's industry produces helicopters, tractors, fertilizers, and other products that would be sources of viable trade, and that trade would indeed be mutually beneficial to both countries.[62] Russia is one of the world's largest countries, with approximately 7 million square kilometers of territory. While Japan, a highly developed country, is very close to Russia—geographically—and possesses a sophisticated economy, Japan cannot supply Russia with products such as bananas or passion fruit. In the case of Manta, Russian representatives observed: "'we produce oil, gas or derivatives. Now the price of these products is very high....But cities can cooperate to find good solutions for both cities.'...This cooperation cannot necessarily pay in dollars [what can be paid for] with bananas, shrimp, passion fruit, seafood, flowers and other products."[63]

Overall, it seems that CCNs are becoming more and more of a viable and effective means of facilitating interstate relations. Engagement premised on a non-military-security basis may be emergent, but such is, nonetheless, becoming more and more integral to the conduct of interstate relations. CCNs are thus creating alternative avenues and forums for productive and mutually beneficial engagement. Fruits, meats, flowers, oil, energy, tourism, shipping, entertainment, humanitarian aid—the list is growing, evolving, and by virtue of the expansive contacts between states is further laying the foundation for cooperative ventures via CCNs.

DOI: 10.1057/9781137308139

Notes

1   Ria-Novosti, "U.S. Okays Russia's Intention to Help Mexico Fight Drug Criminality," February 18, 2010, accessed on 2 April 2010 <http://en.rian.ru/world/20100218/157921859.html>.

2   Sistema Económico Latinoamericano y del Caribe, "Recent Developments in Economic Relations between the Russian Federation and Latin America and the Caribbean: Institutional and Cooperation Mechanisms for Strengthening Relations," Caracas, August 2011, accessed on 22 July 2012 <http://www.sela.org/attach/258/default/Di14_Recent_developments_in_economic_relations_between_Rusia_ALC.pdf>.

3   See *Access My Library* at <http://www.accessmylibrary.com/coms2/summary_0286–19295667_ITM> for a listing of articles detailing Former Presidents Putin and Fox negotiating energy partnerships between Russia and Mexico, respectively.

4   Ria-Novosti, "Russia, Mexico Agree on Direct Flights," February 17, 2010, accessed on 2 April 2010 <http://en.rian.ru/world/ 20100217/157909635.html>.

5   Ria-Novosti, "Mexico's Interjet Signs $650 mln deal for Russia's Superjet Planes," January 17, 2011, accessed on 23 July 2012 <http://en.rian.ru/business/20110117/162176094.html>.

6   El Diario de Hoy, "Rusos y ucranianos se interesan en turismo local: Turistas rusos tienen alto potencial de consumo y de exigencias," 13 de Septiembre de 2010, elsalvador.com, 20 January 2011 <http://www.elsalvador.com/mwedh/nota/ nota_completa.asp?idCat=6374&idArt=5139448>.

7   Ministry Of Foreign Affairs of the Russian Federation, Information and Press Department, "Press Release: Official Visit to Russia of Mexico's Secretary of Foreign Affairs, Patricia Espinosa Cantellano," October 9, 2008, accessed on 2 April 2010 <http://www.ln.mid.ru/brp_4.nsf/e78a48070f128a7b4325699900 5bcbb3/156466a2ccbeb544c32574dd0052c93d?OpenDocument>.

8   Data is this section is drawn from Embassy of the Russian Federation in the United States of Mexico, "Trade and Economic Relations between Russia and Mexico," no date, accessed on 4 April 2010, <http://www.embrumex.com.mx/en_rumex_coop.html>.

9   Ibid.

10 Ibid.

11 Sistema Económico Latinoamericano y del Caribe, "Economic Relations Between the Russian Federation and Latin America and the Caribbean: Current Situation and Prospects," Caracas, Venezuela, July 2009, accessed on 21 July 2012 <http://www.sela.org/attach/258/EDOCS/SRed/2009/07/T023600003569–0-Economic_relations__Russian_Federation_and_ LAC.pdf>.

DOI: 10.1057/9781137308139

12   Ibid.

13   See Ria-Novosti, "Mexican Foreign Secretary to Discuss Trade Relations
     with Russia," June 25, 2011, accessed on 23 July 2012 <http://en.rian.ru/
     world/20110625/164834534.html>.

14   See NY Daily News, "Alfonso Cano, Top FARC Leader," Killed, November
     05, 2011, accessed on 1 November 2012 <http://articles.nydailynews.
     com/2011–11–05/news/30364981_1_guillermo-leon-saenz-alfonso-cano-
     farc>.

15   US Senate Armed Services Committee Hearing, *Current And Future
     Worldwide Threats To The National Security Of The United States,* 111 Congress,
     1 Session, March 10, 2009, US Government Printing Office, Washington
     D.C., 2010, accessed on 31 January 2011 <http://www.gpo.gov/fdsys/pkg/
     CHRG-111shrg54639/html/CHRG-111shrg54639.htm>. See also Victor
     Uribe-Uran, *Colombian Strategic Culture,* October 2009, Florida International
     University-Applied Research Center, Miami, Florida, 31 January 2011< http://
     strategicculture.fiu.edu/Studies.aspx>;

16   Henry Meyer, "Colombia, U.S. Ally, May Conclude Defense Agreement With
     Russia," *Bloomberg.com,* October 3, 2008, accessed on 2 April 2010 <http://
     www.bloomberg.com/apps/news?pid=20601087&sid=atrTYtEVGzbw>.

17   Embassy of the Russian Federation in the Republic of Colombia, "Relaciones
     Ruso-Colombianas," no date, accessed on 3 April 2010 <http://www.
     colombia.mid.ru/rel.html> (translated from the Spanish).

18   Ibid.

19   Ibid.

20   Sistema Económico Latinoamericano y del Caribe, "Recent Developments in
     Economic Relations between the Russian Federation and Latin America and
     the Caribbean: Institutional and Cooperation Mechanisms for Strengthening
     Relations."

21   Ibid.

22   Kirsten Begg, "Colombia to Cultivate Trade Relations with Russia,"
     Colombiareports.com, March 30, 2010, accessed on 28 January 2011 <http://
     colombiareports.com/colombia-news/economy/8926-colombia-to-cultivate-
     relations-with-strategic-partner-russia.html>.

23   Sistema Económico Latinoamericano y del Caribe, "Economic Relations
     Between the Russian Federation and Latin America and the Caribbean:
     Current Situation and Prospects." See Harold Trinkunis, "Venezuelan
     Strategic Culture," July 2009, Florida International University-Applied
     Research Center, Miami, Florida, accessed on 1 February 2011 <http://
     strategicculture.fiu.edu/Studies.aspx>.

24   BBC News, "Venezuela Welcomes Russian Ships," *British Broadcasting
     Corporation,* November 25, 2008, accessed on 2 Feb. 2010 <http://news.bbc.
     co.uk/go/pr/fr/-/2/hi/americas/7747793.stm>.

DOI: 10.1057/9781137308139

25   See Sistema Económico Latinoamericano y del Caribe, "Economic Relations Between the Russian Federation and Latin America and the Caribbean: Current Situation and Prospects."

26   Vladimir Isachenkov, "Chavez: Latin States Should Partner With Russia Against U.S.," *The New York Sun*, September 21, 2008, accessed on 10 February 2010 <http://www.nysun.com/foreign/Chávez-latin-states-should-partner-with-russia/ 86267/ ?print =7256385621>.

27   See El Universal, "Rusia y Venezuela firman acuerdo para extraer crudo en la Faja del Orinoco," February 1, 2010, accessed on 4 April 2010 < http://internacional.eluniversal.com/2010/02/01/eco_ava_rusia-y-venezuela-fi_01A3375653.shtml>.

28   See BBC News Online, "Putin Signs Key Deals with Chavez," April 3, 2010, accessed on 4 April 2010 <http://news.bbc.co.uk/go/pr/fr/-/2/hi/americas/8601388.stm>.

29   Cuban-Russian Relations, no date , accessed on 2 November 2012 <http://www.rtbot.net/Cuba%E2%80%93Russia_relations>.

30   BBC News Online, "Russia to Drill for Oil Off Cuba," July 29, 2009, accessed on 4 April 2010 <http://news.bbc.co.uk/2/hi/ americas/8175704.stm>. See also The Washington Diplomat, "Using Old Friend Cuba as Its Base, Russia Reasserts Its Latin Influence," April 2009, accessed on 2 November 2012 < http://washdiplomat.com/index.php?option=com_content&view=article&id =6257:using-old-friend-cuba-as-its-base-russia-reasserts-its-latin-influence-&catid=976:april-2009&Itemid=257>.

31   Michael Schwirtz, "Cuba and Russia Strengthen Ties as Raúl Castro Visits Moscow," *New York Times Online*, no date, accessed on 3 April 2010 <http://www.nytimes.com/2009/01/30/world/americas/30iht-cuba.4.19816875.html?_r=1>

32   BBC News Online, "Russia to Drill for Oil Off Cuba."

33   U.S. Dept. of State, State Department Documents and Publications, March 29, 2010, Background Notes : Cuba (03/10), Section: State Department Press Release. Lexis-Nexis Academic Universe, 31 March 2010.

34   Ibid.

35   Simon Saradzhyan and Sam Logan, "Russia and Friends in Latin America: Backyard Games," *ISN Security Watch*, October 20, 2008, accessed on 4 February 2010 <http://mexidata.info/id2029.html>.

36   Alwatan News, <www.alwatan.com.kw> December 19, 2008, accessed on 4 April 2010, (trans, from the Arabic, google-docs) <http://translate.google.com/translate?hl=en&sl=ar&u=http://www2.alwatan.com.kw/Default.aspx%3FMgDid%3D705361%26pageId%3D163&ei=2Ai4S5nCoY8ATikcjqAw&sa=X&oi=translate&ct=result&resnum=5&ved=0CBkQ7gEwBA&prev=/search%3Fq%3D%25E2%2580%2598Multipolarity%2Bis%2Ba%2Breality,%2Bwhatever%2Bsupporters%2Bof%2Bthe%2Bunipolar%2Bworld%2Bmight%2Bth

DOI: 10.1057/9781137308139

ink,%2527%2Bthe%2BNicaraguan%2Bpresident%2Bsaid.%26hl%3Den%26sa
fe%3Doff%26client%3Dfirefox-a%26hs%3DRCH%26sa%3DX%26rls%3Dorg.
mozilla:en-US:official%26channel%3Ds>.

37 Ibid.

38 Alexey Pilko, "Nicaragua Canal Project Steps Up to Rival Panama," *Voice
    of Russia Radio*, June 15, 2012, accessed on 24 July 2012 <http://english.ruvr.
    ru/2012_06_15/78160916/>.

39 Alwatan News.

40 Ria-Novosti, "Russia Rows Back on Plans in Nicaragua Channel
    Construction," January 18, 2010, accessed on 3 April 2010 <http://en.rian.ru/
    world/20100118/157601620.html>.

41 Simon Saradzhyan and Sam Logan, "Russia and Friends in Latin America:
    Backyard Games."

42 See Ria-Novosti <http://en.rian.ru/world/20100215/157882370.html>;
    Central American Data, accessed on 1 February 2011 <http://www.
    centralamericadata.com/en/search?q1=country_en_le%3A%22Russia%22>.

43 Dmitry Medvedev, "Congratulations to President of Nicaragua Daniel
    Ortega," January 10, 2012, accessed on 24 July 2012 <http://eng.kremlin.ru/
    news/3312>.

44 See Voice Of America, *News Online*, no date, accessed on 1 April 2010 <http://
    www.voanews.com/english/2008-11-14-voa17.cfm>.

45 El País Online, "Bolivia proyecta comprar armamento ruso por varios
    millones de dólares," El Pais Online, May 22, 2009, accessed on 4 April 2010
    <http://www.elpais.com/articulo/internacional/Bolivia/proyecta/comprar/
    armamento/ruso/varios/millones/dolares/elpepuint/ 20090522elpepuint_10/
    Tes>.

46 BBC News Online, "Russia to Aid Bolivia Drugs Fight," February 17, 2009,
    accessed on 1 April 2010 <http://news.bbc.co.uk/2/hi/americas/7894176.stm>.

47 El País Online, "Bolivia proyecta comprar armamento ruso por varios
    millones de dólares."

48 Ria-Novosti, "Bolivia to Host Servicing Center for Russian Aircraft,"
    October 25, 2009, accessed on 3 April 2010 <http://en.rian.ru/
    world/20091025/156582301.html>.

49 Jean Friedman-Rudovsky, "Casino Gambling: Russia's Export to Latin
    America," *Time/CNN*, December 26, 2009, accessed on 10 February 2010
    <http://www.time.com/time/world/article/0,8599,1945982,00.html>.

50 Ibid.

51 Ibid. Friedman-Rudovsky, "Casino Gambling: Russia's Export to Latin
    America."

52 BBC News Online, "Putin Signs Key Deals with Chávez."

53 See MercoPress, "Latin America Turns to Europe, Russia and China for
    Military Hardware," *MercoPress*, September 9, 2009, accessed on 4 February

DOI: 10.1057/9781137308139

2010 <http://en.mercopress.com/2009/09/08/latinamerica-turns-to-europe-russia-and-china-for-military-hardware>.

54 World Information Service on Energy "New Uranium Mining Projects-South/Central America," *World Information Service on Energy: Uranium Project*, January 27, 2010, accessed on 23 February 2010 <http://www.wise-uranium.org/upsam.html#BRGEN>.

55 BBC News, "Venezuela Welcomes Russian Ships," *British Broadcasting Corporation*, November 25, 2008, accessed on 2 February 2010 <http://news.bbc.co.uk/go/pr/fr/-/2/hi/americas/7747793.stm>.

56 Alex Sánchez, "A COHA Report: Russia Returns to Latin America, Council on Hemispheric Affairs."

57 Russian Federal Customs Service, " Intergovernmental Russian-Ecuador Agreement on Cooperation and Mutual Aid in Customs Affairs, October 29, 2009, accessed on 25 July 2012 <http://eng.customs.ru/index.php?option=com_content&view=article&id=1566:Intergovernmental%20Russian-Ecuador%20Agreement%20on%20Cooperation%20and%20Mutual%20Aid%20in%20Customs%20Affairs%20signed&catid=32:newscat>.

58 Republic of Ecuador Ministry of Foreign Affairs, Trade, and Integration, "Protocol of Information Exchange with the Russian Customs Service," March 16, 2012, accessed on 26 July 2012 <http://www.mmrree.gob.ec/eng/2012/bol0258.asp>.

59 Non-proliferation for Global Security, "Ecuador Plans to Develop Nuclear Plan with Support from Russia," *The Examiner, Caracol TV*, September 3, 2009, accessed on 23 February 2010 <http://npsglobal.org/eng/index.php/news/139-peaceful-uses/713-ecuadorrussia.html>.

60 Reuters, "Russia to Help Ecuador Develop Nuclear Energy," August 21, 2009, 23 February 2010 <http://in.reuters.com/article/oilRpt/idINN2053291620090820?pageNumber=1&virtualBrandChannel=0>.

61 A.E.B.E., "Puerto de Rusia interesado en productos manabitas," April 10, 2008, accessed on 4 April 2010 <http://www.aebe.com.ec/Desktop.aspx?Id=19&art=3311>.

62 Ibid.

63 Ibid.

DOI: 10.1057/9781137308139

5

Concluding Thoughts on CCNs in Russia–Latin America Relations

Abstract: *This work has identified and laid out the nascent and discrete development of complex cooperative networks, and how they are impacting interstate relations. Networks are Russia's primary means of enhancing and diversifying its presence in the region. Russia–Latin America relations illustrate the changes taking place in interstate engagement. Networks appear to be emergent phenomena having a direct impact on how states interact, perceive, and obtain their interests. Trade, investment, and the sundry forms of engagement discussed all point to a significant development within the ongoing dialectic between a system and a society of states; i.e., between the traditional realpolitik model of world order and a cooperative, integrated networked international order, respectively.*

Astrada, Marvin L. and Martín, Félix E. *Russia and Latin America: From Nation-State to Society of States*. New York: Palgrave Macmillan, 2013. DOI: 10.1057/9781137308139.

Since 2008, Russia has maintained extensive and intensive contact and engagement with Latin America, with contacts premised on establishing substantive, long-term networks of cooperation in the realms of security, energy, trade, investment, military affairs, military modernization, resource development, humanitarian aid, and capacity building. After a tour of the region in 2008, which included Venezuela, Cuba, Brazil, and Peru, then-Russian President Medvedev confidently declared that,

> We have visited states that have never been visited either by Russian or Soviet leaders before. This means only one thing: no attention has been paid to these countries. In a sense, we are only just starting fully-fledged, full-format and, I hope, mutually beneficial contacts with the leaders of these states, and with the economies of these states, respectively. There is nothing to feel shy about; one should not fear competition here.[1]

After a long hiatus, Russia has returned to the region, with an ambitious socioeconomic and politico-strategic agenda premised on cultivating soft-power networks of cooperation with the region. For example, Medvedev stressed that, "from the point of view of the energy aspect, the trip was very interesting. Our projects in Venezuela are the most advanced," and in Cuba he and Raúl Castro sought "ways to boost bilateral contacts. The Russian president reiterated that contacts with Latin American [were] humanitarian, economic, energy and defense cooperation."[2]

Russia's ties with Latin America, unlike during the Cold War period, transcend narrow ideological considerations and *realpolitik* concerns based on superpower politics. Russia seeks constructive and complex engagement with a variety of countries, and the region's countries seek the same end. Even those trading partners that pursue *realpolitik* interests, such as Venezuela, have substantive cooperative Complex Cooperative Networks (CCNs) in place. In 2006, for example, a Russian manufacturer of tubes (i.e., TMK) was contracted to expand operations to Venezuela to participate in building a pipe factory, providing Venezuela with the capacity to deliver natural gas to Brazil, Argentina, Uruguay, and Paraguay.[3] The rest of this chapter aims to bring this theme into focus by drawing on the theoretical arguments developed in the first two chapters of this book as well as the empirical evidence presented in the third and fourth chapters.

Soft power networks have thus proven to be Russia's primary means of accomplishing the objective of enhancing and diversifying Russia's

DOI: 10.1057/9781137308139

cooperative partnerships in the region. Russia–Latin America relations appear to confirm that there are indeed serious changes taking place vis-à-vis the means and ends of interstate engagement. CCNs appear to be emergent phenomena that are having a direct say on how states interact, perceive and obtain their interests. Russian engagement with the region is contributing to legitimatizing societal notions of international order. These points of contact are based on societal notions of order such as cooperative non-militarized aspects for interaction and issue of perception and resolution of misunderstanding. Trade, investment, and the various forms of CCN engagements discussed in this work point to a significant development in the ongoing dialectic between a system and a society, i.e., between the traditional distribution-of-power model of world order and the cooperative, integrated networked international order, respectively. Security and defense issues are embedded in Russian engagement. However, Complex Adaptive Systems (CAS)-based CCNs and soft-power resources are not necessarily beholden to or utterly dependant on states for efficacy, viability, and legitimacy vis-à-vis ordering international order and organization.

It is important to note that complexity and CCNs do not produce necessarily stable or predictable systems. The issue of Russian gambling enterprises, as discussed in chapter 4, and the accompanying actors and consequences that follow the gambling industry in Bolivia is indicative of this development. While *realpolitik* concerns remain viable and very important for states, it does not foreclose the potential for CCNs to further evolve and present even more options for engagement and for diversification of state interests beyond military-security interest. As noted in chapter 4, the very notion that Ecuador could engage Russia and seek to develop natural resources and nuclear power capability would have been unthinkable, even ridiculed, during the previous bipolar systemic configuration of international order and organization. While there may be negative results from CCNs and the integration of the global community, this does not completely undermine the potential and possibilities that CCNs have for emerging alternative bases for state interaction.

Seeking to break out from the yoke of tradition, Latin America is actively seeking to diversify and expand its contacts with countries outside of the Americas. These countries share a common desire to shift power from a concentrated to a more diffused configuration. To

DOI: 10.1057/9781137308139

obtain this diffusion, cooperation and networks are being utilized. The politicized "multipolar" notion of international order can be redefined as an attempt to open up the potential of economic development and advancement to more members of the international community, i.e., via multilateral relations that are diverse and networked into a multifaceted plain coinhabited by multiple actors engaged in a multilateral process of international governance.

The Russian foray into Latin America is motivated by different interests. This work has focused on the non-militarized bases for engagement and the non-military-security motives, interests, and goals that inform Russian engagement with the region, and vice versa. In this realm, CCNs assume a critical role in the realization of Russia's and the regions' individual and collective interests. The overarching objective of Russian engagement is premised on providing countries with alternatives, options, and resources when countries consider how best to realize their respective economic interests. By doing so, Russia promotes and develops its own interests from the vantage points of economic diversity, health, and power.

CCNs have not displaced nor abrogated the various realities that continue to define international relations. It would be naïve to think that empirical realities of anarchy and its accoutrements can simply be thought out of existence by the notion and practices of CCNs. Rather, CCNs are emergent and the CAS that is developing in response to a globalizing world order is in the initial phases of laying the foundations for paradigmatic restructuring of global order. For now, CCNs complicate and deepen states' interests and avenues and forums for cooperative engagement. *Realpolitik* is no longer the sole or primary motivating factor or driving force behind states' international behavior. CCNs enable states to look beyond the limiting ordering principles of *realpolitik*, but they have not displaced such principles at this time. Therefore, the claim of this work is that international relations and politics are at present influenced by the coalescing of both *realpolitik* considerations and CCNs' effects. In short, world politics is presently in a transitional condition and flux that requires alternate forms of theory and analysis to make sense of the complex dynamic.

While it is certainly the case that the state remains a viable and preeminent actor on the world stage, systemic changes have been ushered in by the advent of globalization and the rise of CCNs as facilitators. The emergence of complexity has created a global context wherein the

DOI: 10.1057/9781137308139

concentration of power in the state and state-centric accoutrements such as sovereignty, territorial supremacy, infrangible geopolitical borders, the role of supreme regulator, and the singular provider of security, stability, and order, have become diffused. Multifaceted and diverse individuated networks have become an important part of a systemic network that accommodates and thrives on soft as well as hard power precepts. The system of states privileges material power and is committed to anarchy. Anarchy creates severe limits to peace and security in world order. True to his brand of Offensive Realism, John Mearsheimer observes the complex reality of world politics, premised only on selfish considerations and the attendant fear of nation-states living in a self-help system, when he notes that,

> great powers do not work together to promote world order for its own sake, instead each seeks to maximize its own share of world power, which is likely to clash with the goal of creating and sustaining stable international order. This is not to say that great powers never aim to prevent wars and keep the peace. On the contrary, they work hard to deter wars in which they would be the likely victim. In such cases, state behavior is driven largely by narrow calculations about relative power, not a commitment to build a world order independent of a state's own interest.[4]

Within a CAS rubric, global governance becomes more viable, feasible, and conducive to a global communal interest as opposed to adhering to the hyper-selfish character of *realpolitik* state interests. This is the case because states have become hardwired into a systemic context via CCNs. As a CAS, international organization is adapting and learning alternative avenues and forums for engagement and sustained relations that are diverse, that possess breadth and depth beyond the basics of *realpolitik*. The notion that "culturally transmitted directives to action—i.e., social values—exist and are important at each level of system organization, and that value formulations [are] arranged in a hierarchical order to fit the pattern of organizational transaction" plays a role in the imbrication of society onto the states system.[5] This is the case because hyper-connectivity via complex networks creates a context where world actors (states and non-state actors) affect/effect each other.

> It doesn't matter if you know about [globalization effects], and it doesn't matter if you care, they will have their effect anyway. To misunderstand this is to misunderstand the first great lesson of the connected age: we may all have our burdens, but like it or not, we must bear each other's burden as well.[6]

DOI: 10.1057/9781137308139

Growing complexity undergirds every aspect of state behavior—a state's weal, and ultimately its identity—and is ushering modifications to the notion of world order premised on a society of states.

> An unparalleled degree of interconnectivity characterizes daily life at this time particularly for those in the developed world and those in urban settings—whether rich or poor. This interconnectivity is evident in the global economy and in the proliferation of the Internet, voice mail, email, faxes, cell phones, palm pilots, increased air travel, and the ease with which hitherto rare diseases can spread. Our global economy is marked by increased trade, services and investment beyond nation-state borders, and as well, the spread of neo-liberal trade policies. Increasing interconnectedness has also led to the emergence of international governance institutions, international and domestic organizations (NGOs), and transnational corporations. Between 1990 and 2000 alone, the number of NGOs has grown from 6,000 to 26,000. NGOs as a political structure illustrate an emergent property [of a society]. They have the potential of providing an instrument for the growth.[7]

The complex and adaptive dimension of the emergent society utilizes or rather fuses hard and soft power concepts and governance networks in a global venue that accommodates systemic global governance with an incorporated state.

> As the strategic and foreign policies of states are linked to their internal constitutional order, the trade policy of states is another external dimension of the state also connected to the state's internal order. In turn, states must adjust external trade policy objectives and strategies with respect to the international community to accord with their intrinsic political and ideological goals.[8]

As the case of Russia–Latin American relations suggests, complexity is indeed complex; engagement, interaction, networks, and globalism are premised on different types of complexity, such as economic and informational, that, in turn, affect and effect societal as well as states system notions of order and engagement. The complexity that produces and is a product of a CAS creates variegated degrees and levels of engagement among the parts of the system. Hierarchy, connectivity, and dynamism vary according to the degree of integration via CCNs.[9]

The relationship within the larger complex systemic context of world order between state and non-state actors is part of a complex macro-dialectic taking place as far as the location of power in diffuse command

DOI: 10.1057/9781137308139

and control modules. While the state has lost and/or "delegated" its monopoly over the power to define international affairs/reality, it is part of a networked superstructure that blurs the lines between foreign and domestic, that fuses hard and soft power, that becomes part of a more systemic, comprehensive, form of order that is global as opposed to international. Irreconcilable antagonisms between the state and CCNs, or the notion that the state is being "phased out," are inaccurate assessments. Rather, the state is adapting, learning, evolving, and accommodating itself to a complex global environment that is systemically networked. As US Secretary of State Hillary Clinton stated during the recent security conference in Munich, "in today's interconnected world, rapid change is the new norm... [states are] connected with each other and with events around them by technology."[10]

Structural constraints such as anarchy and balance of power are confronted with the rise of a systemic societal notion of governance that is facilitated by CCNs. States and CCNs such as multinational corporations or multinational enterprises cannot simply impose global capitalism and procedural democracy through economic or political coercion, but instead must convince global publics to support these modes of international and domestic order. Cultural, educational, economic, and other forms of soft-power resources and negotiations become operant variables in the systemic super-network that is product and producer of the notion of a "global community." Clinging to a purely system of states power paradigm is, thus, fraught with pitfalls. Zbigniew Brzezinski argues that, in the present age, the US—and by extension, all states—has two choices. The first is to remain locked in a states system paradigm, the cost of which he argues will be devastating and which would lead to "an accelerating plunge into global chaos," while the second choice involves the US supporting the idea of a "global community of shared interest" and "weav[ing] together [with the EU and other CCNs] a broader fabric of binding and institutionalized international cooperation."[11]

Realist theories of international relations tend to dismiss non-state actors as epiphenomenal institutions that are merely tools of implementing the agendas of powerful states. Far from behaving as epiphenomenal tools of the states that create them or as simple forums for coordinating state business—as neo-liberal institutionalist accounts often treat CCNs—CCNs are part of an emergent global society and are assuming pivotal roles in global governance.[12] CCNs "are the beneficiaries of a transfer of state authority in their topical domains. With this

DOI: 10.1057/9781137308139

delegated state authority, [CCNs] issue authoritative decisions and, by virtue of their technical expertise, [CCNs reflect] a normative consensus around common social purposes toward which this expertise might be applied."[13] CCNs and a society of states provide an alternative to the inefficient, violence-ridden, destructive, and war-based system of states in a globalized world. CCNs are more efficient, effective actors, providing venues, linkages, information, knowledge, and drawing upon soft-power resources such as diplomacy, negotiation, data-production, global forums for discussion, logistical support, epistemic communities and expertise, specialization in issue-areas, and cooperation based on mutual (shared) self-interest and common goals. Cooperation provides a more sustainable and stable basis for engagement and interaction. It must be stressed that the emergent CAS is just that—emergent. Therefore, it is neither feasible nor accurate to state that the system of states and a state-centric perspective of world order are antiquated or passé. Rather, the emergent CAS is indicative of an embryonic change taking place in the fabric, contours, and purlieus of the international system, which has been impacted by and is dealing with rising levels of complexity.

Notes

1   Mark A. Smith, "Russia & Latin America: Competition in Washington's 'Near Abroad?'" Paper prepared for Research & Assessment Branch, *Defense Academy of the United Kingdom*, August 9, 2009 (United Kingdom)1–22. See p. 1.

2   Ria-novosti, "Russia Has Returned to Latin America," *Ria-novosti*, November, 28 2008, accessed on 10 February 2010 <http://en.rian.ru/russia/20081128/118588646.html>.

3   elEconomista, "Compañía rusa TMK construirá fábrica de tuberías en Venezuela," *elEconomista.es*, July 28, 2006, accessed on 3 April 2010, <http://www.eleconomista.es/empresas-finanzas/noticias/49408/07/06/Compania-rusa-TMK-construira-fabrica-de-tuberias-en-enezuela.html> (author translation from the Spanish).

4   John Mearsheimer, *The Tragedy of Great Power Politics* (New York: W.W. Norton & Company, 2001), p. 42.

5   Charles McClelland, "General Systems Theory in International Relations." *International Security Systems: Concepts & Models of World Order*, ed. Richard B. Gray (Ithaca: F. E. Peacock, 1969),.p. 31.

DOI: 10.1057/9781137308139

6   Duncan J. Watts, *Six Degrees of Separation: The Science of a Connected Age* (New York: W.W. Norton, 2003), p. 301.
7   Bennett Stark, "A Case Study of Complex Adaptive Systems Theory—Sustainable Global Governance: The Singular Challenge of the Twenty-First Century. University of Ljubljana, WISDOM RISC-Research Paper No. 5, July 2009, pp. 1–38, available online, accessed December 27, 2012, http://www.wisdom.at/Publikation/pdf/RiskBerichte/RRR_BStark_SustainableGlobal_09.pdf, p. 10.
8   Ari Afilalo and Dennis Patterson, "Statecraft, Trade and the Order of States," 6 Chi. J. Int'l L. 725, 728 *Chicago Journal of International Law,* Vol. 6, no. 2 (Winter 2006).
9   Rebecca Dodder and Robert Dare, "Complex Adaptive Systems and Complexity Theory: Inter-related Knowledge Domains," ESD.83: Research Seminar in Engineering Systems, October 31, 2000, Massachusetts Institute of Technology, accessed on February 1, 2011 <http://www.pdfchaser.com/Complex-Adaptive-Systems-and-Complexity-Theory%3A-Inter-related-....html>.
10 Hillary Clinton, quoted by CNN online, February 5, 2011, accessed on 11 February 2011 <http://www.cnn.com/2011/WORLD/europe/02/05/germany.security.conference/index.html?eref=rss_latest&utm_source=feedburner&utm_medium=feed&utm_campaign=Feed%3A+rss%2Fcnn_latest+%28RSS%3A+Most+Recent%29>
11 Zbigniew Brezezinski, *The Choice: Global Domination or Global Leadership* (New York: Basic Books, 2004), p. 218 and p. 219.
12 Rodney Bruce Hall, "Private Authority: Non-state Actors and Global Governance," *Harvard International Review,* June 22, 2005, accessed on 19 April 2010 <http://www.allbusiness.com/public-administration/national-security-international/462542–1.html>.
13 Ibid.

DOI: 10.1057/9781137308139

Bibliography

Books, Book Chapters, and Academic Articles and Reports

Afilalo, Ari and Dennis Patterson. "Statecraft, Trade and the Order of States." *Chicago Journal of International Law*, Vol. 6, No. 2 (Winter 2006), pp. 725–759.

Allison, Graham. "Conceptual Models and the Cuban Missile Crisis." *American Political Science Review*, Vol. 63 (September 1969), pp. 689–718.

▶ Ahmed E., A. S. Elgazzar and A. S. Hegazi. "An Overview of Complex Adaptive Systems." *Mansoura J. Math*, 32 (28 June 2005) 27.0506059 ArXiv (Nonlin).

Astrada, Marvin L. *American Power after 911* (New York: Palgrave Macmillan, 2010).

Astrada, Marvin L. "Strategic Culture: Concept and Application." Applied Research Center, Florida International University, Miami, Florida, February 2010, pp. 5–6, accessed 18 January 2011, http://strategicculture.fiu.edu/Approach/StrategicCultureConceptandApplication.aspx

Bain, Mervyn. Russian-Cuban Relations since 1992: Continuing Camaraderie in a Post-Soviet *World* (Lanham: Lexington, 2008).

Baudot, Jacques. *Building a World Community: Globalization and the Common Good* (Seattle: University of Washington Press, 2001).

Begg, Kirsten. "Colombia to Cultivate Trade Relations with Russia." *Colombiareports.com*, March 30, 2010, accessed 28 January 2011, http://colombiareports.com/

DOI: 10.1057/9781137308139

colombianews/economy/8926-colombia-to-cultivaterelations-with-strategic-partner-russia.html.

Betts, Richard K. *Conflict After the Cold War*, Updated Edition (2nd , ed.) (Longman: New York, 2004).

Bertalanffy, Ludwig von. General System Theory: Foundations, Development, Applications (New York: George Braziller, 1968).

Bertalanffy, Ludwig von. *A Systems View of Man*, ed. Paul A. LaViolette (Boulder: Westview Press, 1981).

Bloom, Howard. "Beyond the Super-computer: Social Groups as Self-Invention Machines." In Albert Somit and Steven A. Peterson (eds), *Sociobiology and Bio-politics. Research in Biopolitics*, Vol. 6 (Greenwich: JAI Press, 1998), pp. 43–64.

Bloom, Howard. Global Brain: The Evolution of Mass Mind from the Big Bang to the 21st Century (John Wiley: New York, 2001).

Bloom, Howard. The Genius of the Beast: A Radical Revision of Capitalism (Amherst: Prometheus, 2010).

Blasier, Cole. "The Soviet Union in Cuban-American Conflict." In Cole Blasier and Carmelo Mesa-Lago (eds) *Cuba in the World* (Pittsburgh: University of Pittsburgh Press, 1979), pp. 37–51.

Bobbitt, Philip. *The Shield of Achilles* (New York: Knopf, 2002).

Bodin, Jean. *On Sovereignty: Four Chapters From the Six Books of the Commonwealth*, , Ed. and trans. Julian H. Franklin (Cambridge: Cambridge University Press 1992).

Bolton, John. "Should We Take Global Governance Seriously?" *Chicago Journal of International Law* Vol. 1, No. 2 (Fall 2000), pp. 205–223.

Brezezinski, Zbigniew. *The Choice: Global Domination or Global Leadership* (New York: Basic   Books, 2004).

Brockner, Eliot. "Russia, Bolivia, and the New US Model." *Latin American Thought*, June 5, 2009, accessed 24 July 2012, http://latamthought. org/2009/06/05/russia-bolivia-and-the-new-us-model/.

Bull, Hedley. *The Anarchical Society: A Study of Order in World Politics*. Foreword by Stanley Hoffmann, 2nd ed. (New York: Columbia University Press, 1995).

Bull, Hedley. *The Anarchical Society: A Study of Order in World Politics*. Foreword by Andrew Hurrell and Stanley Hoffman, 3rd ed. (Colombia University Press: New York, 2002).

Buzan, Barry. "From International System to International Society of States: Structural Realism and Regime Theory Meet the English School." *International Organization*, Vol. 47, No. 3 (1993), pp. 327–352.

DOI: 10.1057/9781137308139

Buzan, Barry. "The Levels of Analysis Problem in International Relations Reconsidered." In Ken Booth and Steve Smith (eds), *International Relations Theory Today* (University Park: Penn State University Press, 1995), pp. 198–216.

Castañeda, Jorge G. "Latin America's Left Turn." *Foreign Affairs*. May/June 2006, accessed 1 April 2010, http://www.foreignaffairs.com/articles/61702/jorge-g-castaneda/latin-americas-left-turn .

Cesarano, Filippo. *Monetary Theory and Bretton Woods: The Construction of an International Monetary Order*, Series: Historical Perspectives on Modern Economics (New York: Cambridge University Press, 2006).

Claude, Inis. *Power and International Relations* (New York: Random House, 1962).

Crampton, R. J. The Hollou Détente: Anglo-German Relations in the Balkans 1911–1914 (Atlantic Highlands: Humanities Press, 1980).

Croucher, Sheila L. Globalization and Belonging: The Politics of Identity in a Changing World (Lanham: Rowman & Littlefield, 2004).

Dash, Robert C. "Globalization for Whom and for What." *Latin American Perspectives*, Vol. 25, No. 6 (1998), pp. 52–54.

Dawkins, Richard. *The Selfish Gene* (New York: Oxford University Press, 1976).

Deese, David A. World Trade Politics: Power, Principles, and Leadership (New York: Routledge, 2008).

Dehousse, Renaud. "Regulation by Networks in the EU: The Role of European Agencies." *Journal of European Public Policy*, Vol. 4 (1997), pp. 246–261.

Deleuze, Gilles and Felix Guattari. "Kafka: Toward a Minor Literature." In Vincent B. Leitch, William E. Cain, Laurie A. Finke, Barbara E. Johnson, John McGowan and Jeffrey J. Williams (eds), trans. Dana Poland, *The Norton Anthology of Theory & Criticism* (New York: WW Norton, 2001).

Dodder, Rebecca and Robert Dare. "Complex Adaptive Systems and Complexity Theory: Inter-related Knowledge Domains," ESD.83: *Research Seminar in Engineering Systems*, October 31, 2000, Massachusetts Institute of Technology, accessed on 1 February 2011 http://www.pdfchaser.com/Complex-Adaptive-Systems-and-Complexity-Theory%3A-Inter-related-....html.

Domínguez, Jorge I. "The Armed Forces and Foreign Relations." In Cole Blasier and Carmelo Mesa-Lago (eds), *Cuba in the World* (Pittsburgh: University of Pittsburgh Press, 1979), pp. 53–86.

DOI: 10.1057/9781137308139

Dooley, Kevin. "A Nominal Definition of Complex Adaptive Systems." *The Chaos Network*, Vol. 8, No. 1 (1996), pp. 2–3.

Dooley, Kevin J. "A Complex Adaptive Systems Model of Organization Change." *Nonlinear Dynamics, Psychology, and Life Sciences*, Vol. 1 (Jan. 1997), pp. 69–97.

Durfee, Edmund, Lesser, Victor, and Corkill, Daniel "Trends in Cooperative Distributed Problem Solving." *IEEE Transactions on Knowledge and Data Engineering*, Volume KDE-1, No. 1 (March 1989), pp. 63–83.

Egan, Daniel and Levon A. Chorbajian (eds) *Power: A Critical Reader* (Upper Saddle River: Pearson, 2005).

Eidelson, Roy J. "Complex Adaptive Systems in the Behavioral and Social Sciences," *Review of General Psychology*, Vol. 1, No. 1 (1997), pp. 42–71.

Fischer, Edward F. Cultural Logics and Global Economies: Maya Identity in Thought and Practice (Austin: University of Texas Press, 2001).

Fischer, Edward F. "Summary Analysis: Guatemala Strategic Culture," June 1, 2010 (unpublished manuscript). Summary prepared for Florida International University.

Fiss, Peer C. and Paul M. Hirsch. "The Discourse of Globalization: Framing and Sense-Making of an Emerging Concept." *American Sociological Review*, Vol. 70 (2005), p. 42.

Flores-Mendez, Roberto A. "Towards the Standardization of Multi-Agent System Architectures: An Overview." *ACM Crossroads, Special Issue on Intelligent Agents*, Association for Computer Machinery, No. 5.4, pp. 18–24 (Summer 1999), (also available online, pp.1–12, see pp. 4, 23) accessed 1 December 2012.

Foucault, Michel. *Archaeology of Knowledge and Discourse of Language*, trans. A.M. Sheridan Smith (New York: Pantheon, 1972).

Franke, Katherine M. "The Domesticated Liberty of Lawrence v. Texas." *Columbia Law Review* Vol. 104, No. 5 (2004), pp. 1399–1426.

Friedman, Thomas L. *Longitudes and Attitudes*, 2002, accessed 4 January 2011, http://www.thomaslfriedman.com/longitudes prologue.htm.

Gray, Richard B. International Security Systems: Concepts & Models of World Order (Ithaca: F. E. Peacock, 1969).

Giddens, Anthony. *The Consequences of Modernity* (Cambridge: Polity Press, 1990).

Generation Online. "Jean Bodin: On Sovereignty," no date, accessed on 12 January 2011, http://www.generation-online .org/p/fpbodin1.htm .

DOI: 10.1057/9781137308139

González, Edward. "Institutionalization, Political Elites, and Foreign Policies." In Cole Blasier and Carmelo Mesa-Lago (eds), *Cuba in the World* (Pittsburgh: University of Pittsburgh Press, 1979), pp. 3–37.

Govind Gokhale. *Balkrishna Asoka Maurya* (Twaynes: New York, 1966).

Guarnizo, Luis Eduardo and Michael Peter Smith. "The Locations of Trans-nationalism." *Comparative Urban and Community Research*, no date, p. 2, accessed on 19 January 2011, hcd.ucdavis.edu/faculty/.../smith/.../Locations_of_transnationalism.pdf

Haas, Ernst. Where Knowledge is Power: Three Models of Change in International Organizations (Berkeley: University of California Press, 1990).

Hall, J. A. *States in History* (New York: Blackwell, 1986).

Hall, Rodney Bruce. "Private Authority: Non-state Actors and Global Governance." *Harvard International Review*, June 22, 2005, accessed on 19 April 2010, http://www.allbusiness.com/public-administration/national-securityinternational/462542-1.html.

Held, David, Tony McGrew, Jonathan Perraton, and David Goldblatt. *Global Transformations* (Cambridge: Polity, 1999).

Hinsley, F. H. *Sovereignty*, 2nd ed. (Cambridge: Cambridge University Press, 1986).

Hobbes, Thomas. *The Leviathan*. Ed. Michael Oakesshott, (New York: Collier, 1962).

Hobsbawm, Eric J. *Nations and Nationalism Since 1780* (New York: Cambridge University Press, 1990).

Holland, John H. "Complex Adaptive Systems." *Daedalus,* Vol. 121, No. 1 (Winter 1992), pp. 17–30.

Hollis, Martin and Steve Smith. *Explaining and Understanding International Relations* (New York: Oxford University Press, 1990).

Huntington, Samuel. *The Third Wave: Democratization in the Late Twentieth Century* (Norman: University of Oklahoma Press, 1991).

Jennings, Nicholas. R., Katia Sycara, and M. Wooldridge. "A Roadmap of Agent Research and Development." In Jennings, Nicholas R., Katia Sycara and Michael Georgeff (eds), *Autonomous Agents and Multi-Agent Systems Journal*, Vol. 1, No.1, pp. 7–38 (Kluwer Academic Publishers: Boston, 1998).

Jenson, Jane and Boaventura de Sousa Santos. "Introduction: Case Studies and Common Trends in Globalization." In Jane Jenson and Boaventura de Sousa Santos (eds), *Globalizing Institutions; Case Studies in Regulation and Innovation* (Aldershot: Ashgate, 2000), pp. 9–29.

DOI: 10.1057/9781137308139

Jervis, Robert. *Perception and Misperception in International Politics* (Princeton: Princeton University Press, 1976).

Jervis, Robert. *System Effects: Complexity in Political and Social Life* (Princeton: Princeton University Press, 1997).

Kant, Immanuel. *Perpetual Peace: A Philosophical Essay.* Trans. and intro. M. Campbell Smith, preface by L. Latta (London: George Allen and Unwin, 1917).

Kaplan, Morton. System and Process in International Politics (New York: Wiley, 1957).

Karns, Margaret P., and Karen A. Mingst. 2010. *International Organizations: The Politics and Processes of Global Governance.* Boulder, Colorado: Lynne Rienner Publishers.

Keck, Margaret E. and Kathryn Sikkink. *Activists Beyond Borders: Advocacy Networks in International Politics* (Ithaca: Cornell University Press, 1998).

Keohane, Robert O. "Governance in a Partially Globalized World." Presidential Address, Annual Meeting of the APSA, 2000. *American Political Science Review*, Vol. 95, No. 1 (March 2001), pp. 1 –13.

Keohane, Robert O. and Joseph S. Nye, Jr. "Trans-Governmental Relations and International Organizations." *World Politics* Vol. 27, No. 1 (October 1974), pp. 39–62.

Keohane, Robert O. and Joseph S. Nye, Jr. *Power and Interdependence: World Politics in Transition* (Boston: Little, Brown, 1977).

Kiely, Ray. The Clash of Globalizations: Neo-Liberalism, the Third Way and Anti-Globalization (Haymarket: New York, 2009).

Laszlo, Ervin. The Systems View of the World: The Natural Philosophy of the New Developments in the Sciences (New York: George Braziller, 1972).

Laszlo, Ervin. The Systems View of the World: A Holistic Vision for Our Time Advances in Systems Theory, Complexity, and the Human Sciences (Cresskill: Hampton Press, 1996).

Luban, David. "A Theory of Crimes Against Humanity." *Yale Journal of International Law*, Vol. 29 (2004), pp. 85–167.

Lyotard, Jean Francois. *The Postmodern Condition: A Report on Knowledge.* Foreword by Fredric Jameson, trans. Geoff Bennington and Brian Massumi (Minneapolis: University of Minnesota Press, 1984).

Lyotard, Jean Francois and Jean-Loup Thebaud. *Just Gaming*, trans. Wlad Godzich (Minneapolis: University of Minnesota Press, 1999).

DOI: 10.1057/9781137308139

Martín, Félix E. Militarist Peace in South America: Conditions for War and Peace (New York: Palgrave Macmillan, 2006).

MacCormick, Neil. "Beyond the Sovereign State." *Modern Law Review*, Vol. 56, No. 1 (January 1993), pp. 1–18.

Mandelbrot, Benoît. "How Long Is the Coast of Britain? Statistical Self-Similarity and Fractional Dimension Science." *New Series*, Vol. 156, No. 3775 (May 5, 1967), pp. 636–638.

Marx, Karl and Frederick Engels. *The Communist Manifesto, A Modern Edition* (New York: Verso, 1998).

Maturana, Humberto R. "The Organization of the Living: A Theory of the Living Organization." *International Journal of Man-Machine Studies* Vol. 7, No. 3 (May 1975), 313–332.

Mbembe, Achille. *On the Postcolony* (Berkley: University of California Press, 2001).

McClelland, Charles. "General Systems Theory in International Relations." *International Security Systems: Concepts & Models of World Order*, ed. Richard B. Gray (Ithaca: F. E. Peacock, 1969).

McClory, Jonathan. *The New Persuaders II*. Institute for Government, 2011, accessed 26 July 2012, http://www.instituteforgovernment.org.uk/sites/default/files/publications/The%20New%20PersuadersII_0.pdf

Mearsheimer, John. *The Tragedy of Great Power Politics* (New York, W.W. Norton & Company, 2001).

Mendras, Marie. "Soviet Policy Toward the Third World." Soviet Foreign Policy, *Proceedings of the Academy of Political Science*, Vol. 36, No. 4 (1987), pp. 164–175.

Moul, William. "The Level of Analysis Problem Revisited." *Canadian Journal of Political Science*, Vol. 6, No. 3 (September 1973), pp. 494–513.

Naím, Moisés. "Five Wars of Globalization." *Foreign Policy*, Vol. 134 (January/February 2003), pp. 29–36.

Nietzsche, Friedrich. "On Truth & Lying in a Non-Moral Sense." In Vincent B. Leitch, William E. Cain, Laurie A. Finke, Barbara E. Johnson, John McGowan and Jeffrey J. Williams (eds), trans. Ronald Spiers, *The Norton Anthology of Theory & Criticism* (New York: WW Norton, 2001), pp. 874–884.

Nye, Jr., Joseph S. "Neorealism and Neoliberalism." *World Politics*, Vol. 40 (1988), pp. 235–51.

Nye, Jr., Joseph S. *Bound to Lead: The Changing Nature of American Power* (Cambridge: Harvard University Press, 1990).

DOI: 10.1057/9781137308139

Nye, Jr., Joseph S. *Soft Power: The Means to Success in World Politics* (New York: Public Affairs, 2004).

Nye, Jr., Joseph S. "Public Diplomacy and Soft Power." *Annals of the American Academy of Political and Social Science*, Vol. 616 (March 2008), pp. 94–109.

Onuf, Nicholas. "Sovereignty: Outline of a Conceptual History." *Alternatives*, Vol. 16, No.4 (1991), pp. 425–446.

Osiander, Andreas. "Sovereignty, International Relations, and the Westphalian Myth." *International Organization*, Vol. 55, No. 2 (2001), pp. 251–287.

Petras, James and Henry Veltmeyer. "Globalization or Imperialism?" In Daniel Egan and Levon A. Chorbajian (eds) *Power: A Critical Reader* (Upper Saddle River: Pearson, 2005).

Piccotto, Sol. "Networks in International Economic Integration." *Northwestern Journal of Law and Business*, Vol. 17 (1996–1997), p. 1014.

Plato. "The Laws, Book 1 Selections." *The Human Condition: Philosophical Issues War and Peace. The Philosophy Resource Center*, accessed 3 March 2010 http://www.radicalacademy.com/hcwpfilehome5a.htm

Poncela, J., J. Gómez-Gardeñes, L. M. Floría, A. Sánchez, and Y. Moreno. "Complex Cooperative Networks from Evolutionary Preferential Attachment." *Plos One*, Vol., 3, No. 6 (2008), pp. e2449, accessed on 23 December 2012, http://www.plosone.org/article/info:doi/10.1371/journal.pone.0002449

Ray, James L. *Global Politics*, 2nd ed. (Boston: Houghton Mifflin, 1983).

Reinicke, Wolfgang H. "Global Public Policy." *Foreign Affairs*. Vol. 76 (1997), p. 137.

Rosenau, James, N. (ed.) *The Scientific Study of Foreign Policy*, revised , (New York: Nichols Publishing, 1980).

Rosenau, James, N. *Turbulence in World Politics* (Princeton: Princeton University Press, 1990).

Rossi, Ino. Frontiers of Globalization Research: Theoretical and Methodological Approaches (New York: Springer, 2007).

Ruggie, John Gerard. Constructing the World Polity: Essays on International Institutionalization   (New York: Routledge, 1998).

Shapiro, Michael J. Methods and Nations: Cultural Governance and the Indigenous Subject (New York: Routledge, 2004).

Shapiro, Martin. "Administrative Law Unbounded: Reflections on Government and Governance." *Indiana Journal of Global Legal Studies*, Vol. 8, No. 2 (2001), pp. 369–377, available online, accessed

DOI: 10.1057/9781137308139

27 December 2012 http://www.repository.law.indiana.edu/ijgls/vol8/
iss2/6.

Simon, H. A. "Near Decomposability and Complexity: How a Mind
Resides in a Brain," Santa Fe Institute Studies in the Sciences of
Complexity, Proceedings Vol. XXII. In H. Morowitz and J. L. Singer
(eds), *The Mind, the Brain, and Complex Adaptive Systems* (Reading:
Addison Wesley, 1995), pp. 25–43.

Singer, J. D. "The Level of Analysis Problem in International Relations."
World Politics, Vol. 14, No. 1 (October 1961), pp. 77–92.

Singer, Marshall R. Weak States in a World of Powers: The Dynamics of
International Relationships (New York: Macmillan, 1972).

Slaughter, Anne-Marie. *A New World Order* (Princeton: Princeton
University Press, 2004).

Smith, Mark A. "Russia & Latin America: Competition in Washington's
'Near Abroad?'" Paper Prepared for Research & Assessment Branch,
Defense Academy of the United Kingdom, UK, pp. 1–22, August 9,
2009, available online, accessed 24 December 2012, http://mercury.
ethz.ch/serviceengine/Files/ISN/.../09_Aug.pdf.

Snyder, Jack and Robert Jervis, (eds) *Coping With Complexity in the
International System* (Boulder: Westview Press, 1993).

Stanford Encyclopedia of Philosophy, "Jean Bodin," March 25, 2005,
revised June 14, 2010, accessed 12 January 2011, http://plato.stanford.
edu/entries/bodin/#4 .

Stark, Bennett. A Case Study of Complex Adaptive Systems
Theory—Sustainable Global Governance: The Singular Challenge
of the Twenty-First Century. University of Ljubljana, WISDOM
RISC-Research Paper No. 5, July 2009, pp. 1–38, available online,
accessed December 27, 2012, http://www.wisdom.at/Publikation/pdf/
RiskBerichte/RRR_BStark_SustainableGlobal_09.pdf

Stayer, Joseph R. *On the Medieval Origins of the Modern State* (Princeton:
Princeton University Press, 1970).

Thomson, Janice. "State Sovereignty in International Relations: Bridging
the Gap between Theory and Empirical Research." *International
Studies Quarterly*, Vol. 39, No.2 (1995), pp. 213–233.

Tolstoy, Leo. *War and Peace.* Trans. Anthony Briggs, intro. Orlando Figes
(New York: Penguin Books, 2006).

Trinkunis, Harold. "Venezuelan Strategic Culture." Applied Research
Center, Florida International University, July 2009, Miami, Florida,
accessed 1 February 2011 http://strategicculture.fiu.edu/Studies.aspx .

DOI: 10.1057/9781137308139

Turku, Helga. Isolationist States in an Interdependent World (Burlington: Ashgate, 2009).

Uribe-Uran, Víctor. "Colombian Strategic Culture." Applied Research Center, Florida International University, October 2009, Miami, Florida, accessed 31 January 2011, http://strategicculture.fiu.edu/Studies.aspx

Varela, Francisco J. "Autonomy and Autopoiesis." In Gerhard Roth and Helmut Schwegler (eds), *Self-Organizing Systems: An Interdisciplinary Approach* (New York: Campus Verlag, 1981).

Waldrop, M. Mitchell. Complexity: The Emerging Science at the Edge of Order & Chaos (New York: Penguin, 1992).

Waltz, Kenneth. *Theory of International Politics* (Reading, Massachusetts: Addison-Wesley, 1979).

Watts, Duncan J. Six Degrees of Separation: The Science of a Connected Age (New York: W.W. Norton, 2003).

Wendt, Alexander. *Social Theory of International Politics* (Cambridge: Cambridge University Press, 1999).

Whitehead, Alfred North. *Symbolism: It's Meaning & Effect* (New York: Fordham University Press, 1927).

Wittgenstein, Ludwig. *Tractatus Logico-Philosophicus*. Trans. D. F. Pears and B.F. McGuiness. (London: Routledge, 1975).

World Commission on Culture and Development (WCCD), *Our Creative Diversity* (Paris: UNESCO 1995).

Governmental and NGOs' Documents and Reports

Banathy, Bela. "A Taste of Systemics." The Primer Project: A Special Integration Group (SIG) of the International Society for the Systems Sciences (ISSS) (originally SGSR, Society for General Systems Research) and the International Institute for Systemic Inquiry and Integration: The First International Electronic Seminar on Wholeness, December 1, 1996 to December 31, 1997, accessed on 25 February 2008, http://www.newciv.org/ISSS_Primer/seminar.html.

Blank, Stephen. "Russia in Latin America: Geopolitical Games in the U.S.'s Neighborhood." Paper prepared for IFRI Russia/NIS Center, April 2009, accessed 25 March 2010, www.ifri.org/.../ifri_Blank_Russia_and_LatinAmerica_ENG_April_09.pdf.

DOI: 10.1057/9781137308139

Brown Lee, Jason. "Complex Adaptive Systems." Technical Report: Complex Intelligent Systems Laboratory, Centre for Information Technology Research, Faculty of Information Communication Technology, Swinburne University of Technology, Melbourne, March 2007, accessed on 21 July 2012, http://www.scribd.com/doc/22947963/Complex-Adaptive-Systems

Dooley, Kevin J. 2010. "Complexity in the Social Science Glossary. A Research Training Project of the European Commission." At http://www.irit.fr/COSI/glossary/fulllist.php?letter=M, accessed 2 April 2010.

Embassy of the Russian Federation in the Republic of Colombia. "Relaciones Ruso-Colombianas," no date, accessed on 3 April 2010, http://www.colombia.mid.ru/rel.html.

Embassy of the Russian Federation in the United States of Mexico, "Trade and Economic Relations between Russia and Mexico," no date, accessed on 4 April 2010, <http://www.embrumex.com.mx/en_rumex_coop.html>.

Institute for the Future, "The Cooperation Project: Objectives, Accomplishments, And Proposals." March 30, 2005, accessed on 20 May 2010 www.rheingold.com/cooperation/CooperationProject_3_30_05.pdf.

Lavrov, Sergey. "The New Stage of Development of Russian-Latin American Relations," *Ministry of Foreign Affairs of the Russian Federation*, August 24, 2011, accessed on 22 July 2012, http://www.mid.ru/brp_4.nsf/0/A27D6F235094016DC32578F70042C31C.

McClory, Jonathan. *The New Persuaders: An International Ranking of Soft Power*. Institute for Government, December 2010, accessed on 21 July 2012, http://www.scribd.com/doc/47790532/THE-NEW-PERSUADERS-An-international-ranking-of-soft-power .

Ministry of Foreign Affairs of the Russian Federation, Information and Press Department. "Press Release: Official Visit to Russia of Mexico's Secretary of Foreign Affairs, Patricia Espinosa Cantellano," October 9, 2008, accessed on 2 April 2010, http://www.ln.mid.ru/brp_4.nsf/e78a48070f128a7b43256999005bcbb3/156466a2ccbeb544c32574dd0052c93d?OpenDocument.

Ministry of Foreign Affairs of the Russian Federation. "On the Meeting of the Minister of Foreign Affairs of Russia S.V. Lavrov and the Minister of Foreign Affairs of Cuba Bruno Rodríguez," July 11, 2012, accessed on 27 July 2012, http://www.mid.ru/bdomp/brp_4.nsf/e78a4

DOI: 10.1057/9781137308139

8070f128a7b43256999005bcbb3/1ddbd4d4188cc78e44257a390024ced7
!OpenDocument.

Ministry of Foreign Affairs of the Russian Federation. "About the
Official Visit Minister of Foreign Affairs of Peru R. Roncayolo to
Russia," May 29, 2012, accessed on 31 July 2012, http://www.mid.ru/
bdomp/brp_4.nsf/e78a48070f128a7b43256999005bcbb3/33566c5d54c7
47e444257a1600384990!OpenDocument.

Republic of Ecuador Ministry of Foreign Affairs, Trade, and Integration.
"Protocol of Information Exchange with the Russian Customs
Service," March 16, 2012, accessed on 26 July 2012, http://www.
mmrree.gob.ec/eng/2012/bolo258.asp.

Russian Federal Customs Service. " Intergovernmental Russian-Ecuador
Agreement on Cooperation and Mutual Aid in Customs Affairs,
October 29, 2009, accessed on 25 July 2012, http://eng.customs.
ru/index.php?option=com_content&view=article&id=1566:Int
ergovernmental%20RussianEcuador%20Agreement%20on%20
Cooperation%20and%20Mutual%20Aid%20in%20Customs%20
Affairs%20signed&catid=32:news-cat.

Sánchez, Alex. "A COHA Report: Russia Returns to Latin America,
Council on Hemispheric Affairs," 2009, accessed 27 March 2010,
http://www.printfriendly.com/print?url=http%3A%2F%2Fwww.coha.
org%2Frussia-returns-to-latin america%2F&partner=sociable.

Sistema Económico Latinoamericano y del Caribe. "Economic
Relations Between the Russian Federation and Latin America and
the Caribbean: Current Situation and Prospects," Caracas, Venezuela,
July 2009, accessed on 21 July 2012, http://www.sela.org/attach/258/
EDOCS/SRed/2009/07/T023600003569–0-Economic_relations__
Russian_Federation_and_ LAC.pdf.

Sistema Económico Latinoamericano y del Caribe. "Final Report
of the Regional Meeting on Recent Developments in Economic
Relations Between the Russian Federation and Latin America and
the Caribbean," Caracas, Venezuela, May 21, 2012, accessed on 22 July
2012, http://www.sela.org/attach/258/default/Final_Report_Recent_
Developments_in_Economic_relations_between_the_Russian_LAC.
pdf.

Sistema Económico Latinoamericano y del Caribe. "Recent
Developments In Economic Relations Between the Russian
Federation and Latin America and the Caribbean: Institutional and
Cooperation Mechanisms for Strengthening Relations," Caracas,

DOI: 10.1057/9781137308139

Venezuela, August 2011, accessed on 22 July 2012, http://www.sela.
    org/attach/258/default/Di14_Recent_developments_in_economic_
    relations_between_Rusia_ALC.pdf .
United States Department of Defense. *Dictionary of Military Terms.*
    Accessed 15 July 2009, http://www.dtic.mil/doctrine/jel/doddict/.
United States Library of Congress. "Country Study: Soviet Union." U.S.
    Library of Congress, Federal Research Division, May 1989, accessed
    on 18 Feb. 2010, http://lcweb2.loc.gov/cgi-bin/query/r?frd/cstdy:@
    field%28DOCID+su0269%29.
United States Senate Armed Services Committee Hearing. *Current and
    Future Worldwide Threats to the National Security of the United States*, 111
    Congress, 1 Session, March 10, 2009, US Government Printing Office,
    Washington D.C., 2010, accessed on 31 January 2011, http://www.gpo.
    gov/fdsys/pkg/CHRG-111shrg54639/html/CHRG111shrg54639.htm.
UNRISD. States of Disarray: The Social Effects of Globalization
    (Geneva: UNRISD, 1995).
Walle, Walter. "Russia Turns to the South for Military and Economic
    Alliances." Council on Hemispheric Affairs, May 8, 2012, accessed
    on 22 July 2012, http://www.coha.org/russia-turns-to-the-south-for-
    military-and-economic-alliances/ .
World Commission on Culture and Development. *Our Creative
    Diversity.* (Paris, France: UNESCO, 1995).
World Information Service on Energy. "New Uranium Mining Projects-
    South/Central America." *World Information Service on Energy:
    Uranium Project*, January 27, 2010, accessed on 23 February 2010,
    http://www.wise-uranium.org/upsam.html#BRGEN.

Newspaper Articles and Reports

A.E.B.E., "Puerto de Rusia interesado en productos manabitas," April
    10, 2008, accessed on 4 April 2010, http://www.aebe.com.ec/Desktop.
    aspx?Id=19&art=3311.
Alwatan News at www.alwatan.com.kw. December 19, 2008, accessed on
    4 April 2010, trans, from the Arabic by google-docs: http://translate.
    google.com/translate?hl=en&sl=ar&u=http://www2.alwatan.com.
    kw/Default.aspx%3FMgDid%3D705361%26pageId%3D163&ei=2Ai4S
    5nCoY8ATikcjqAw&sa=X&oi=translate&ct=result&resnum=5&ved
    =0CBkQ7gEwBA&prev=/search%3Fq%3D%25E2%2580%2598Multip

DOI: 10.1057/9781137308139

olarity%2Bis%2Ba%2Breality,%2Bwhatever%2Bsupporters%2Bof%2B
the%2Bunipolar%2Bworld%2Bmight%2Bthink,%2527%2Bthe%2BNic
araguan%2Bpresident%2Bsaid.%26hl%3Den%26safe%3Doff%26clien
t%3Dfirefoxa%26hs%3DRCH%26sa%3DX%26rls%3Dorg.mozilla:en-
US:official%26channel%3Ds.

Bancroft-Hinchey, Timothy. "Russia Boosts Relations with Latin
America." *Pravda.RU*, April 5, 2010, accessed on 20 January 2011,
http://english.pravda.ru/world/americas/05–04–2010/112853-
russia_latin_america-0/#.

BBC News Online. "Venezuela Welcomes Russian Ships." November
25, 2008, accessed on 2 February 2010, http://news.bbc.co.uk/go/pr/
fr/-/2/hi/americas/7747793.stm.

BBC News Online. "Russia to Aid Bolivia Drugs Fight." February
17, 2009, accessed on 1 April 2010, http://news.bbc.co.uk/2/hi/
americas/7894176.stm.

BBC News Online. "Russia to Drill for Oil Off Cuba." July 29,
2009, accessed on 4 April 2010, http://news.bbc.co.uk/ 2/hi/
americas/8175704.stm .

BBC News Online. "Putin Signs Key Deals with Chávez." April 3, 2010,
accessed on 4 April 2010, http://news.bbc.co.uk/go/pr/fr/-/2/hi/
americas/8601388.stm.

Clinton, Hillary. *CNN online*, February 5, 2011, accessed 11 February 2011,
http://www.cnn.com/2011/WORLD/europe/02/05/germany.security.
conference/index.html?eref=rss_latest&utm_source=feedburner&utm_
medium=feed&utm_campaign=Feed%3A+rss%2Fcnn_latest+%28RSS
%3A+Most+Recent%29.

Chomsky, Noam. "Noam Chomsky Chats with Washington Post
Readers." *The Washington Post,* March 24, 2006, accessed on 11 April
2008, http:// www.chomsky.info/debates/20060324.htm.

Cuban-Russian Relations. No date, accessed on 2 November 2012,
http://www.rtbot.net/Cuba%E2%80%93Russia_relations.

DiploNews. "Russian-Cuban Relations Are 110 Years Strong." July
20, 2012, accessed on 31 July 2012, http://www.diplonews.com/
articles/2012/20120720_RussiaCuba.php.

El Diario de Hoy. "Rusos y ucranianos se interesan en turismo local:
Turistas rusos tienen alto potencial de consumo y de exigencias,"
13 de Septiembre de 2010, *elsalvador.com*, accessed 20 January
2011, http://www.elsalvador.com/mwedh/nota/nota_completa.
asp?idCat=6374&idArt=5139448

DOI: 10.1057/9781137308139

El Economista. "Compañía rusa TMK construirá fábrica de tuberías en Venezuela." *elEconomista.es*, July 28, 2006, accessed on 3 April 2010, http://www.eleconomista.es/empresasfinanzas/noticias/49408/07/06/Companiarusa-TMK-construira-fabrica-de-tuberias -en-Venezuela.html.

El País Online. "Bolivia proyecta comprar armamento ruso por varios millones de dólares." *Elpaís.com*, May 22, 2009, accessed on 4 April 2010, http://www.elpais.com/articulo/internacional/Bolivia/proyecta/comprar/armamento/ruso/varios/millones/dolares/elpepuint/20090522elpepuint_10/Tes.

El Universal. "Rusia y Venezuela firman acuerdo para extraer crudo en la Faja del Orinoco." February 1, 2010, accessed on 4 April 2010, http://internacional.eluniversal.com/2010/02/01/eco_ava_rusia-y-venezuela-fi_01A3375653.shtml.

El Universo.com. "Economía Rusia y Dinamarca inauguran línea transatlántica entre Rusia y Ecuador." El Universo.com, Guayaquil, Ecuador, March 22, 2010, accessed on 26 January. 2011, ttp://www.eluniverso.com/2010/03/22/1/1356/rusia-dinamarca-inauguran-linea-transatlantica-carga-rusia-ecuador.html.

Friedman-Rudovsky, Jean. "Casino Gambling: Russia's Export to Latin America." *Time/CNN Online*, December 26, 2009, accessed on 10 February 2010, ttp://www.time.com/time/world/article/0,8599,1945982,00.html.

González, Eduardo José. "Russia and Latin American Strengthen Judicial Cooperation." *Radiohc*, March 30, 2012, accessed on 28 July 2012, http://www.radiohc.cu/ing/news/cuba/6210-russia-and-latin-american-strengthen-judicial-cooperation.html.

Isachenkov, Vladimir. "Chávez: Latin States Should Partner With Russia Against U.S." *The New York Sun*, September 21, 2008, accessed 10 February 2010, http://www.nysun.com/foreign/Chávez-latin-states-should-partner-with-russia/ 86267/ ?print =7256385621.

Meyer, Henry. "Colombia, U.S. Ally, May Conclude Defense Agreement with Russia." Bloomberg.com, October 3, 2008, accessed on 2 April 2010, http://www.bloomberg.com/apps/news?pid=20601087&sid=atrTYtEVGzbw.

Monocle Magazine. "A Briefing on Global Affairs, Business, Culture & Design: Soft Power Survey." December 2010/January 2011, Vol. 4, no. 39, pp. 41–50, available online at http://stage.monocle.com/magazine/issues/39/the-new-soft-sell/.

DOI: 10.1057/9781137308139

Montealegre, Oscar. "Brazil's Trade Diplomacy." *Diplomatic Courier*, January 20, 2011, accessed on 30 July 2012, http://www.diplomaticourier.com/news/bric/37.

Medvedev, Dmitry. "Congratulations to President of Nicaragua Daniel Ortega." January 10, 2012, accessed on 24 July 2012, http://eng.kremlin.ru/news/3312.

MercoPress. "Latin America turns to Europe, Russia and China for Military Hardware." *MercoPress*, September 9, 2009, accessed on 4 February 2010, http://en.mercopress.com/2009/09/08/latinamerica-turns-to-europe-russia-and-china-for-military-hardware.

New York Times. "Free Gaza Movement's Clash with the State of Israel in Isabel Kershner, "Deadly Israeli Raid Draws Condemnation." *New York Times Online*, May 31, 2010, accessed on 9 June 2010, http://www.nytimes.com/2010/06/01/world/middleeast/ 01flotilla.html.

New York Daily News. "Alfonso Cano, Top FARC Leader, Killed." November 5, 2011, accessed on 1 November 2012, http://articles.nydailynews.com/2011–11–05/news/30364981_1_guillermo-leon-saenz-alfonso-cano-farc.

Nicholson, Alex and Andre Soliani, "Russia, Brazil Plan to Buy $20 Billion IMF Bonds," June 10, 2009, accessed on 1 April 2010 <http://www.bloomberg.com/apps/news?pid=20670001&sid=a5nc3eTSovTc>.

Nikandrov, Nil. "Russia – Latin America: The Union of Solidarity and Pragmatism." *Ria-novosti*, June 21, 2010, accessed on 23 July 2012, http://en.rian.ru/international_affairs/20100621/159513144.html.

Nonproliferation for Global Security. "Ecuador Plans to Develop Nuclear Plan with Support from Russia." *The Examiner*, Caracol TV, September 3, 2009, accessed on 23 February 2010, shttp://npsglobal.org/eng/index.php/news/139-peaceful-uses/713-ecuadorrussia.html.

Pilko, Alexey. "Nicaragua Canal Project Steps Up to Rival Panama." *Voice of Russia Radio*, June 15, 2012, accessed on 24 July 2012, http://english.ruvr.ru/2012_06_15/78160916/.

Pravda.Ru. "Russia Offers Latin America to Combat Drugs Together." March 12, 2012, accessed on 22 July 2012,http://english.pravda.ru/russia/economics/12032012/120748russia_latin_america_drugso/#.http://lcweb2.loc.gov/cgi-bin/query/r?frd/cstdy:@field%28DOCID+su0269%29.

Pravda.Ru. "Guatemala Wants Russian Arms in Exchange for Coffee and Sugar." March 29, 2010, accessed on 29 March 2010, http://english.pravda.ru/world/americas/23–03–2010/112681-guatemala-0.

DOI: 10.1057/9781137308139

Putin, Vladimir. "Speech, 43rd Munich Security Conference, 10 Feb. 2007." BBC News Online, accessed on 1 June 2007, http://news.bbc. co.uk/2/hi/europe/6349287.stm.

Reuters. "Russia to help Ecuador Develop Nuclear Energy." August 21, 2009, accessed on 23 February 2010, http://in.reuters.com/article/oilRpt/idINN2053291620090820?pageNumber=1&virtualBrandChan nel=0.

Ria-Novosti. "Russia Has Returned to Latin America." November 28, 2008, accessed on 10 February 2010, http://en.rian.ru/russia/20081128/118588646.html.

Ria-Novosti. "U.S. Okays Russia's Intention to Help Mexico Fight Drug Criminality." February 18, 2010, accessed on 2 April 2010, http://en.rian.ru/world/20100218/157921859.html .

Ria-Novosti. "Russia, Mexico Agree on Direct Flights." February 17, 2010, accessed on 2 April 2010, http://en.rian.ru/world/20100217/157909635.html.

Ria-Novosti. "Mexico's Interjet Signs $650 mln Deal for Russia's Superjet Planes." January 17, 2011, accessed on 23 July 2012, http://en.rian.ru/business/20110117/162176094.html.

Ria-Novosti. "Mexican Foreign Secretary to Discuss Trade Relations with Russia." June 25, 2011, accessed on 23 July 2012, http://en.rian.ru/world/20110625/164834534.html.

Ria-Novosti. "Russia Rows Back on Plans in Nicaragua Channel Construction." January 18, 2010, accessed on 3April 2010 http://en.rian.ru/world/20100118/157601620.html.

Ria-Novosti. "Bolivia to Host Servicing Center for Russian Aircraft." October 25, 2009, accessed on 3 April 2010, http://en.rian.ru/world/20091025/156582301.html .

Ria-Novosti. Central American Data Set. No date, accessed on 1 February 2011, http://www.centralamericadata.com/en/search?q1=country_en_le%3A%22Russia%22 . http://en.rian.ru/world/20100215/157882370.html.

Saradzhyan, Simon and Sam Logan. "Russia and Friends in Latin America: Backyard Games." *ISN Security Watch*, October 20, 2008, accessed on 4 February 2010, http://mexidata.info/id2029.html.

Schwirtz, Michael. "Cuba and Russia Strengthen Ties as Raúl Castro Visits Moscow." *New York Times Online*, 2009, accessed on 3 April 2010, http://www.nytimes.com/2009/01/30/world/americas/30iht-cuba.4.19816875.html?_r=1.

DOI: 10.1057/9781137308139

Tumgazeteler. "A Look at Russian-Latin American Relations as 'New Cold War' Talks Gain Momentum." September 26, 2008, accessed 10 February 2010, http://www.tumgazeteler.com/?a=4157397.

The Washington Diplomat. "Using Old Friend Cuba as Its Base, Russia Reasserts Its Latin Influence." April 2009, accessed 2 November 2012, http://washdiplomat.com/index.php?option=com_content&view=article&id=6257:using-old-friend-cuba-as-its-base-russia-reasserts-its-latin-influence-&catid=976:april-2009&Itemid=257.

Voice of America. *News Online*, no date, accessed on 1 April 2010, http://www.voanews.com/english/2008–1–14-voa17.cfm .

Zeenews.com. "Russia Seeks to Boost Ties with Latin America." February 3, 2010. At http://www.zeenews.com/printstory.aspx?id=601165, accessed 10 February 2010.

DOI: 10.1057/9781137308139

Index

Adaptive learning 19, 36, 40
Agency 12, 18–21, 33, 93
Agent 14, 16–21, 41, 44, 50
Anarchy 2–4, 17–20, 30, 38–39,
  41–42, 44, 104–107
Argentina 31–32, 38, 61, 68,
  70–71, 91, 102

Bodin, Jean 3, 45–46
Bloom, Howard 16, 35, 45, 48
Bolivia 32, 38, 61, 69, 71, 76–77,
  90–91, 94
Brazil 29, 31–32, 38, 43, 61,
  68–71, 76–77, 92–93, 102
Bretton Woods 6, 30

Castro, Fidel 61, 64–66, 71, 85
Castro, Raul 71, 85–87, 102
Chavez, Hugo 61, 81
Cold War 12, 30,62–63, 67, 70,
  72, 84–87, 94, 102
Colombia 33, 38, 59, 68, 71–72,
  76–77, 81–84, 89–91, 94
Complex Adaptive System 1,
  5–6, 11–12, 14–21, 30, 37–43,
  48–50
Complex Cooperative
  Networks 1, 4–5, 12, 18,
  21, 29, 41, 44, 70, 72, 76,
  103–104, 106, 108
Complex interdependence 30,
  37, 43
Complexity 4–6, 11, 18, 22, 27, 41,
  44, 66, 68, 71, 97, 99, 101, 103

Correa, Rafael 32, 61, 94
Cuba 17, 29, 31–32, 38, 61, 64–67,
  70–71, 76–77, 84–88, 102
Cultural logics 21

Diversification 44

Ecuador 32, 61, 69, 72, 76–77,
  82, 92–95, 103
Energy 13, 26, 29, 31–32, 38, 44,
  59–60, 68, 71–72, 77–79, 81,
  83–85, 87–92, 94–95, 102

FARC (Revolutionary Armed
  Forces of Colombia) 81–82
Friedman, Thomas 4

Gambling 91, 103
General Systems Theory 14–15,
Globalism 4, 7, 19, 106
Globalist 4, 6, 17, 44–45, 47, 50,
  70–71
Globalization 3–5, 7, 14, 26–28,
  30, 32, 41, 45, 48–49, 62,
  104–105

Hard power 8, 11, 15, 27, 61, 69,
  81, 105
Holland, John 16, 41

Ideology 28, 60, 61–66, 70,
  86–88, 94
International
  organization 3–15, 18–20,

DOI: 10.1057/9781137308139

26, 34, 37–39, 42–45, 49, 58, 61, 67, 71, 76–77, 91, 105

Investment 12–13, 22, 26, 28–29, 38, 43–44, 59–60, 65–66, 68, 70–71, 77–81, 83, 86–88, 90–93, 101–103, 106

Lavrov, Sergei 68, 70, 72, 93

Meme 37, 41–42, 48
Mexico 33, 38, 71, 76, 78–81, 84, 86, 91
Morales, Evo 32, 61, 90–91
Multilateral 7, 59, 61, 66, 72, 79, 82–84, 88, 93, 104,
Multipolar 34, 59–60, 69, 79, 82, 84–85, 88, 90, 104
Munich Security Conference 69, 107

Narco-trafficking 78, 82–83, 89, 91
Networks 1, 4–15, 21–22, 26–49, 57–60, 67, 69–71, 76–78, 81–87, 91–95, 101–106
Nicaragua 33, 61, 66–70, 76–77, 87–90
Nuremberg Tribunal 5–6, 40

Ortega, Daniel 33, 61, 88

Peace of Westphalia 2–3, 27
Peru 29, 32, 61, 66–67, 71, 91–93, 102
Plato 2–3

Pluri-polar 34, 71–72, 82
Putin, Vladimir 68–73

Realism 3, 105
Realpolitik 6, 10, 21, 32, 38, 46, 58, 60, 63, 67, 72, 77, 81–84, 88–94, 101–105

Self-similarity 18
Society of states 6–8, 11–12, 18–20, 26–28, 31, 44–45, 47–50, 72, 76–77, 101, 106–108
Soft power 11
Sovereignty 5, 18–19, 21, 27, 44–47, 105
System of states 2–7, 10–11, 18–20, 27, 29–36, 42–47, 57–58, 62, 67–69, 77, 88, 105–108

Terrorism 7, 32, 40, 66, 79, 81–82, 93
Tourism 59, 71, 79–80, 83, 95

U.S.S.R. 12, 61–69, 80, 86, 94
Unipolar 34, 69, 71
United Nations 5–6, 10, 17, 32

Venezuela 29, 31–34, 38, 59, 61, 64, 68, 71, 76–77, 81–85, 88, 102

Waltz, Kenneth 3, 15
Wendt, Alexander 19

DOI: 10.1057/9781137308139

Lightning Source UK Ltd.
Milton Keynes UK
UKOW04n0611270913

218055UK00002B/7/P